Knowledge in the Time of Cholera

KNOWLEDGE *in the* TIME OF CHOLERA

THE STRUGGLE OVER AMERICAN MEDICINE IN THE NINETEENTH CENTURY

OWEN WHOOLEY

THE UNIVERSITY OF CHICAGO PRESS
Chicago and London

OWEN WHOOLEY is assistant professor of sociology at the University of New Mexico.

The University of Chicago Press, Chicago 60637
The University of Chicago Press, Ltd., London

Printed in the United States of America

22 21 20 19 18 17 16 15 14 13   1 2 3 4 5

ISBN-13: 978-0-226-01746-4 (cloth)
ISBN-13: 978-0-226-01763-1 (paper)
ISBN-13: 978-0-226-01777-8 (e-book)

Parts of chapters 1 and 2 were published in "Organization Formation as Epistemic Practice: The Early Epistemological Function of the American Medical Association," *Qualitative Sociology* 33, no. 4 (2011): 491–511. Reprinted with kind permission from Springer Science+Business Media.

Library of Congress Cataloging-in-Publication Data

Whooley, Owen, author.
Knowledge in the time of cholera : the struggle over American medicine in the nineteenth century / Owen Whooley.
pages cm
Includes bibliographical references and index.
ISBN 978-0-226-01746-4 (cloth : alkaline paper) — ISBN 978-0-226-01763-1 (paperback : alkaline paper) — ISBN 978-0-226-01777-8 (e-book) 1. Cholera—United States—History—19th century. 2. Medicine—United States—History—19th century. 3. Knowledge, Sociology of. I. Title.
RC131.A2W46 2013
614.5'14097309034—dc23

2012036982

♾ This paper meets the requirements of ANSI/NISO Z39.48-1992 (Permanence of Paper).

To my mom, Candie

A conflict of ideas is a battle or series of battles, and presents all the meannesses, the cunning, the stratagems, the bitterness and sometimes the violence of actual war.

WILLIAM H. HOLCOMBE

President of the American Institute of Homeopathy, 1874 to 1876

CONTENTS

ACKNOWLEDGMENTS

THE GREAT FICTION OF INTELLECTUAL LIFE IS THAT books usually bear a single name. All inquiry is communal, and all insight, interactive. This book is the sum of innumerable conversations, real or imagined, that opened my eyes, prodded me forward, and ultimately gave me great pleasure. More so than the final product, it is these conversations that I treasure.

For six years, I called the Department of Sociology at New York University my home. Jeff Goodwin constantly pushed me to keep the big picture in mind. Historical sociologists always run the risk of getting lost in the trees; Jeff helped me see the forest. Craig Calhoun provided me with generous support over my graduate career—so generous that I will never be able to adequately express my appreciation. Always forthcoming in offering me the type of penetrating feedback for which he is well known, he has an uncanny ability to distill key points into a form digestible and understandable to all. Most of the more articulate points in this book can be traced back to him. Ann Morning has a calming presence that is contagious. She is a model of professionalism I can only hope to be able to shamelessly ape one day. To Troy Duster goes the credit of the most useful advice I received during this process—write, write, and then write some more. By reducing this project to the simple exercise of writing, Troy gave me the strategy by which I persevered. Ed Lehman, my reader, has been a constant mentor and friend. Vivek Chibber, David Garland, and Richard Sennett all at some point have read my work and provided invaluable feedback. Finally, I would like to thank the often overlooked administrative staff, who humored my pestering throughout the years, especially Dominick Bagnato, Candyce Golis, and Jamie Lloyd.

To have dear friends who are also your teachers is to be blessed. Jane Jones is the only person to have read every single word of every single piece of my research. She is the consummate editor; without her, my verbosity runs wild. I can never thank her enough for everything. Noah McClain, my fellow chronicler of the absurd, has been a steadfast source of support and much needed merriment. I cultivated my sociological imagination with him, as we subjected the most mundane and ridiculous topics to its wrath, over beers of course. Claudio Benzecry has given me so much advice and enjoyable conversation that I can only hope to someday return even half of it. I would also like to thank (and apologize to) that merry bunch in my writing workshop for enduring my long drafts with smiles of encouragement: Hannah Jones, Sarah Kaufman, Amy LeClair, Tey Meadow, Ashley Mears, Harel Shapira, and Grace Yukich. A special thank-you to Gabi Abend, Rene Almeling, and Andrew Deener, who helped see this book to fruition with prompt and penetrating feedback. Finally, I'd like to thank everyone who participated in NYLON with me, that cauldron of creative sociology, especially Ruthie Braunstein, Ernesto Castaneda, Monika Krause, David Madden, Erin O'Connor, Marion Wrenn, and Mark Treskon, who deserves special mention for constructing the fancy licensing map for the book.

Through the various accidents of my meandering biography, I have been fortunate enough to be a part of not one, but four excellent academic communities. I spent my first two years of graduate school in the Sociology Department at Boston College, and it was there that I figured out who I wanted to be intellectually. Bill Gamson, Char Ryan, and Diane Vaughan are the models of generous scholars. As a NIMH postgraduate fellow in the Institute of Health, Health Care Policy and Aging Research, I was granted the two most important ingredients for successfully writing a book—abundant time and access to interested minds. I'd like to thank Allan Horwitz, David Mechanic, Deborah Carr, Gerry Grob, Eviatar Zerubavel, and my fellow fellows Ken MacLeish, Tyson Smith, and Zöe Wool. My current home is the Sociology Department of the University of New Mexico, where I look forward to years of sharing rarefied conversations with my new colleagues in the rarefied air of Albuquerque.

This book would not have been possible without the archivists and librarians who were more than willing to help me find that elusive document, despite my violations of certain library rules. Toiling away in anonymity, it is they who preserve our past, and for this, they should be celebrated. The staff at the New York Academy of Medicine, particularly Arlene Shaner, deserves much of my gratitude as we spent the better part of two years together. I'd also like to thank the staffs at the Bobst Library at New York University, the

Bradford Homeopathic Collection at the Taubman Health Science Library at the University of Michigan, the Butler Library at Columbia University, the New York History Society, the Parnassus Library at the University of California—San Francisco, and the Rockefeller Foundation Archive Center. Historical sociologists depend on the diligent research of historians who keep us honest and make sure we get the details of the story straight. I stood on the shoulders of many historian/giants, whose contributions deserve more recognition than mere references. Of note has been the scholarship of Harris Coulter, John Duffy, John S. Haller Jr., Howard Markel, Charles Rosenberg, and John Harley Warner. Finally, I'm especially grateful to editor—and keen wit—Doug Mitchell, both for his encouragement and his sharp repartee; I am humbled that my book is now a small part of the indelible mark he has made on the discipline of sociology. Also at the University of Chicago Press, I'd like to thank Tim McGovern, Jennifer Rappaport, and Carol Saller.

Research, especially historical research, can breed insularity and can reward socially deviant behaviors. Fortunately, I am blessed with a rich community of loved ones, who couldn't care less about the ivory tower and who keep me grounded. My dad, Jim, was a dreaming empiricist and playful rationalist if ever there was one. It was he who taught me to approach the world with humble curiosity. While his death antedates this book, my writing is, and will always be, a conversation with him. My brother and best friend, Michael, is a source of constant laughs. He grounds me in who I am, and I couldn't imagine experiencing any of life's vicissitudes without him right next to me. My office mate and walking partner, Jibbs, is a constant source of delight. My closest friends—Sam Fillian, Karyn Miller, Elizabeth Pfifer Payne, and Mary Regan—are the most genuine people I have ever met. Erin Tarica, my wife and my word, is probably the only person to find my bookishness attractive, even generously misrepresenting it as "cool." Her exuberance draws me out; her intuitive sensitivity is the perfect antidote to my excessive rationality. Words cannot express my feelings; hopefully, a lifetime of laughter will. Finally, this book is dedicated to my mom, Candie, who has been my greatest advocate and sturdiest support throughout my life. To merely thank her is inadequate; I am who I am because of her.

INTRODUCTION

Of Cholera, Quacks, and Competing Medical Visions

When cholera—a disease never seen nor imagined by American physicians—first arrived in the United States in 1832, the horrors it unleashed belied any pretensions doctors may have held to medical mastery. With its dramatic symptoms and rapid mortality, the foreign disease created widespread panic, which intensified as physicians' interventions did nothing to stem its progress. Worse still, physicians' heroic therapies, like bloodletting, sped the disease's course, undermining the ability of cholera patients to stave off death.

Cholera's ability to "mock the calculations of man" (Short quoted in Chambers 1938, 164) generated a crisis in medicine as doctors scrambled impotently to find a cure for the unfamiliar disease. Their failure to do so seriously compromised the public standing of the medical profession. By the 1830s, allopathic or regular physicians—that is, the dominant sect of physicians later represented by the American Medical Association (AMA)[1]—had gained a measure of professional control, as thirteen state legislatures had passed medical licensing laws (Numbers 1988). These laws represented early cultural validation, a crucial step in the consolidation of professional authority. Cholera destroyed this momentum. Alternative medical movements pointed to allopathy's failures during the cholera epidemic to mount a campaign to repeal the licensing laws. And by the mid-1840s, merely a decade after they had been passed, the licensing laws were universally repealed (the one exception being the New Jersey statute). Cholera became a symbolic failure for allopathic medicine that ushered in an era of unregulated medicine and intense competition among medical sects.

Contrast this reaction with a later foreign epidemic. In 1918, the "Spanish" influenza hit the United States for the first time. The "great influenza" epidemic was the worst in the country's history (Barry 2005), and by all objective measures, an even more stunning failure for the medical profession. The influenza claimed between 500,000 and 670,000 Americans and upward of five million worldwide, 3 percent of the world's population. By contrast, the 1832 cholera epidemic killed only about 10,000 Americans. Yet, unlike in 1832, no one questioned the authority of the allopathic medical profession. Fortified within institutions like hospitals, laboratories, and research universities, regulars, through the AMA, had achieved such a high level of professional authority that they remained unchallenged by even the most deadly of epidemics. Operating under the banner of science and the bacteriological paradigm, allopathic physicians had wrested control of medicine from competing sects, winning recognition as *the* experts in treating, controlling, and understanding disease. The medical profession emerged from the influenza epidemic as influential as ever.

Two epidemics, two conspicuous failures, and yet two widely divergent professional outcomes. What happened to the profession in the intervening years, between these two epidemic bookends, that these failures could yield such different results? Cholera destabilized allopathic authority and led to a retraction of professional privileges that gave birth to a period of rancorous medical competition; the Spanish influenza reinforced physicians' clout as local, state, and federal agencies turned to them—and only them—to combat the flu (Barry 2005). These contrasting outcomes reflected a dramatic shift in the allocation of professional authority between the two epidemics—a shift whose origin clearly was not driven by the profession's effectiveness in combating infectious diseases. The medical profession failed in both cases, and by all standards, much more spectacularly in 1918. And yet the ramifications of these failures differed greatly. Why was the public unwilling to extend the benefit of the doubt to allopathic doctors during the deadly cholera, but willing to during the deadlier influenza epidemic?

Over the same period, medicine underwent a major change in epistemology, in how medical knowledge was produced and understood. During the 1832 cholera epidemic, Dr. Henry Bronson, a professor at Yale Medical School, offered this account of the disease:

> If I am asked the essential, non-contagious cause of cholera, I answer frankly—*I do not know*. Every agent in nature, real or imaginary, has been accused. Electricity, magnetism, earth, air, water, sun, moon, planets, comets, have each been arraigned in vain. There is a mystery which hangs over the origin and spread of epidemics, which will probably never be re-

moved. The philosophers of the present day are no wiser on this subject than those who lived three thousand years ago. (1832, 86)

Typical of the writings on cholera during the first two cholera epidemics, Bronson's muddled assessment reveals an unstable definition of the disease. Many possible causes, from comets to magnetism, are interrogated; all are found wanting. Bronson betrays the great confusion of allopathy, not only toward the cause of cholera, but also where to even look for such a cause. There was no paradigmatic account of disease. It is not that Bronson cannot make sense of the evidence; he has little vision of what the evidence should even look like. Absent this, he throws out some ideas and, finally dejected, expresses skepticism that cholera's mystery will ever be solved. This is the lament of a doctor, who not only lacks a coherent road map to make sense of the disease, but doubts that such a road map even exists. His despondency is palpable. This inchoate account, however, does gesture toward the dominant perception of medical epistemology during the 1832 cholera epidemic. Lamenting that the "*philosophers of the present day* are no wiser on this subject than those who lived three thousand years ago," Bronson identifies medicine as a practice of philosophy. This reflects the prevailing view of medicine during the period. Medicine was an exercise in philosophizing rather than scientific researching. Deploying rationalist, speculative systems to understand disease, doctors were less focused on interpreting facts; they were in the business of crafting elaborate philosophical systems.

Sixty years later, William Osler offered a very different account of cholera. In *The Principles and Practices of Medicine*, the preeminent medical textbook of the period, Osler (1895, 132) concisely defines cholera as "specific, infectious disease, caused by the comma bacillus of Koch, and characterized clinically by violent purging and rapid collapse." Here we have a much neater definition of cholera. Gone are the uncertainty, the hesitating prose, and the horde of causal candidates. Cholera is now a specific disease caused by a specific microorganism. The authority appealed to is Robert Koch, a German laboratory scientist. In offering a scientific fact, plainly and succinctly stated, Osler displays confidence in the epistemological foundation of his profession. No longer armchair philosophers, Osler's doctors derive hard truths from the laboratory. Medicine is a science.

Bronson and Osler would hardly recognize the cholera of which the other speaks. The idea of cholera being caused by a microscopic organism would have seemed just as absurd to Bronson, as searching for cholera's etiology in comets would have for Osler. This incommensurability speaks to a radical shift in medical epistemology. Typically, epistemological change is viewed as the straightforward product of scientific advancement, the progressive illumination of truth. Such accounts, however, invert temporality; epistemological systems must be accepted prior to recognizing something as truth, as these systems set the standards by which truth is judged. Bronson to Osler was not just a shift from ignorance to insight, from darkness to light. It involved a reformulation of what medical knowledge *is* and how it is to be obtained. To accept cholera as a germ meant accepting the laboratory as the loci of medical insight and the disease cultures growing in these labs as legitimate medical facts. For Bronson and his peers, this would have been unthinkable. But by 1895, they were well on their way to becoming medical common sense. In sixty years, medical knowledge had made the dramatic leap from an exercise in philosophical inquiry to a science rooted in the experimental methods of the laboratory.

These two changes—a dramatic shift in professional authority on one hand and an epistemological change on the other—coalesced to produce a medical profession unusual in the developed world. These changes were not just concurrent but intimately related, each contributing to the development of a medical profession widely recognized as exceptional among developed countries, in its steadfast—and successful—opposition to government incursions into medical practice (most evident in the AMA's various campaigns against government-run health insurance), and its fixation on scientific and technological solutions. While physicians of many different countries eventually embraced the laboratory sciences, none did so with as much fervor as the U.S. medical profession.

In this book, I explain how debates over medical professionalization in the nineteenth century—conflicts over issues like licensing, board of health composition, government recognition of alternative medical sects—*became* epistemological, in the sense that underlying these specific issues was the animating question of what constituted legitimate medical knowledge. Professional struggle was inextricably tied to fundamental intellectual debates waged on the level of epistemology, in which the very identity of what constituted a medical fact was at stake. Or more accurately, these professional debates were *made* epistemological through a confluence of broad social changes, which enabled epidemics like cholera, and alternative medical movements, which seized the opportunities afforded by cholera to force allopathic medicine to justify its expertise in epistemological terms. Cholera and quacks joined to foment epistemological angst for allopathic medicine. And the eventual character of the profession would be inscribed with their indelible marks, as the allopaths would have to solve their riddles to achieve professional authority.

By emphasizing epistemological struggle and change, I provide a framework to better understand the professionalization of U.S. medicine and, in doing so, offer a more nuanced account of a key (if not *the* key) cause of the exceptionalism of the U.S. medical system. When professionalization is no longer viewed as flowing directly from medical discoveries, and instead is seen as evolving according to the vicissitudes of epistemic politics, a very different story of the professionalization of American medicine emerges. It is a story of missed opportunities, of intellectual roads not taken, and of significant contributions by alternative medical movements that have all but disappeared from our historical consciousness. It is a story of the recurrent failures of allopathic physicians to reconcile their professional aspirations with the democratic ideals of American culture, and the repeated acceptance and recognition of alternative movements by government institutions. It is a story of the consolidation of professional authority by allopathic physicians through a strategy that circumvented the state—and public oversight—by securing the financial support of private philanthropies. But foremost, it is a story about the long-standing tension between the logic of professionalization and the ideals of democracy out of which an exceptional U.S. medical profession was born—one that raises difficult questions about the role of professions in democratic cultures.

HEROIC DISCOVERIES, MISLEADING STORIES

Conventional accounts of U.S. medical professionalization link professional authority to improvements in medical knowledge without paying attention to the epistemological changes that underlie these "improvements." In these narratives, the allocation of authority follows on the heels of scientific discoveries, properly meted out according to the merit of the knowledge attained. We can call these accounts, somewhat crudely, "truth-wins-out narratives." The logic of these narratives, often referred to as the "diffusion model" (Latour 1987), holds that ideas, by their self-evident truth, force people to assent to them. Advocates of these discoveries are subsequently accorded requisite acclaim and authority, often in the form of professional privileges.

In the case of the professionalization of U.S. medicine, the truth-wins-out narrative has two variations. In its most basic form, professional authority was awarded to those who subscribed to true ideas, who made the key "discoveries" in the new science of bacteriology. "True" ideas won out over lesser, partial truths. Professionalization followed the imperatives of scientific progress, with the consolidation of professional power achieved through the gradual, rational incorporation of scientific principles into medical practice (see, for example, Bulloch 1979; de Kruif 1996; Duffy 1993; Rutkow 2010). The second variation of the truth-wins-out narrative shares the logic of scientific progress but adds the element of efficacy. Here professional authority is achieved, not only through the power of ideas, but in the results they obtain, as developments in medical knowledge led to more effective therapeutic or sanitary interventions, which legitimated professional authority of physicians. The germ theory rapidly led to new cures that justified its assumptions and solidified the authority of doctors.

The seductive common sense of the truth-wins-out narrative muddies how we understand the history of the U.S. medical profession, the progress in medicine generally, and the origins of the tremendous authority afforded doctors in the United States. It ignores the heterogeneity of nineteenth-century medical thought, reducing professionalization to a single linear process. Rather than exploring the ascendancy of one *way* of thinking about medicine, it tells of a single march toward progress in medical knowledge. It obscures significant ideas once entertained but ultimately discarded and dismisses alternative medical movements as aberrations or mere repositories of ignorance, that is, if they are even considered at all. Intellectual developments are decontextualized, depoliticized, and presented as evolving according to the dictates of disinterested research rather than as part of a specific professional project. And the dissemination of bacteriological ideas is given no explanation other than a vague appeal to the propulsion of their self-evident truth: "By this time [1880] the news of Koch's discoveries had *spread* to all of the laboratories of Europe and *had crossed the ocean* and inflamed the doctors of America" (de Kruif 1996 [1926], 119). Exercising what E. P. Thompson (1968, 12) terms the "condescension of posterity," the truth-wins-out narrative naturalizes professional authority, transforming the flux of the nineteenth century into something of a predetermined outcome.

While the truth-wins-out narrative is rarely laid out as explicitly as this, its underlying logic persists in our understandings of professionalization, as it squares with commonsense notions of the development of science, notions which the sociology of science has spent decades combating. But the history of knowledge does not conform to the strictures of common sense, and the truth-wins-out narrative cannot bear the weight of historical scrutiny. Take the two key discoveries in the history of cholera lauded in the truth-wins-out narrative—John Snow's discovery of the waterborne nature of the disease in 1855 and Robert Koch's identification of the cholera microbe in 1884. These true ideas, rapidly accepted, gave birth to bacteriology and scientific medicine, which led to the disease's demise. A tidy story, yes, but it's wrong. Snow's famous cholera map can only be considered a tipping point in the debate over the etiology of cholera by reading history backward. The map, almost an afterthought in Snow's work, had little effect in convincing skeptics of the validity of the contagion theory (Koch 2005; Vinten-Johansen et al. 2003). There is almost no mention of it in American allopathic journals prior to the twentieth century. Likewise, Koch's widely reported discovery of *Vibrio cholerae* did not provide the decisive "win" for the bacteriological model of cholera (Rosenkrantz 1985; Warner 1991), as it was beset with inconsistencies that fostered widespread skepticism (Rothstein 1992, 267). And in terms of combating cholera, the bacteriological model did not produce much in the way of improvements in therapeutics (unlike diphtheria or rabies, no widely used cholera vaccine was ever embraced) or prevention (effective sanitary improvements were done in the name of the now discredited miasmic theory of disease)[2] (Dubos 1987; Duffy 1990; McKeown 1976, 1979). These issues—the ambiguity surrounding the theory initially and the lag between the promise of the germ theory and its results—are not just evident in the history of cholera; generally, the biomedical model only yielded significant therapeutic advantages in the 1930s, long after it was accepted by allopaths as legitimate (Spink 1978). Thus, in 1892, the year of the final U.S. cholera epidemic and the dawn of allopathic professional control, the efficacy of the bacteriological model existed largely in its promise. This messy, ambiguous historical record of cholera thus begs the questions, how did this disease come to be seen as a microbe and how did this understanding get folded into the professional project of allopathic medicine?

Despite these problems, the truth-wins-out narrative has proven obstinately resilient, even as historians have challenged it on a number of grounds (see Grob 2002; Warner 1997, 1998). Were it restricted to publications of the American Medical Association (AMA) or the myths doctors tell themselves, it might not be much of a concern. The problem is that its assumptions insinuate themselves into more critical sociological analyses of professionalization. As such, it is not enough to dismiss it as merely "hagiographic" (Warner 1997, 2).

It is not surprising that older functionalist accounts of professionalization embrace the truth-wins-out logic. When professions are viewed as arising to fulfill some preexisting societal need or structural imperative (see Parsons 1964), it is difficult to maintain the critical distance necessary to challenge the science that justifies such a role. But what of the more critical sociological research that arose in opposition to functionalism? These critical accounts depict professionalization not as a functional response to a societal need but rather as a political process that involves winning allies and creating a strong organizational infrastructure to promote professional goals.[3] Still, even this research inadvertently reproduces a version of truth-wins-out logic. Here the issue is reproduction through neglect. In reacting against the truth-wins-out narrative, which gives undo power to ideas, critical analyses tend to ignore ideas altogether, mustering organizational and political explanations for the professionalization of medicine (i.e., Berlant 1975; Freidson 1970, 1988; Larson 1977). Through this silence, they unintentionally reproduce misguided assumptions of the truth-wins-out narrative; focused on the *organizational* infrastructure of professions, they neglect the *intellectual* infrastructure. Ideas come to serve merely as window dressing for the real politicking happening behind the scenes.

The power of the truth-wins-out logic is displayed in the way it insinuates itself into the preeminent sociological treatment of the U.S. medical profession, Paul Starr's *The Social Transformation of American Medicine* (1982). Critical of both functionalism and purely organizational accounts of professionalization, Starr seeks to integrate organizational and cultural factors, recognizing professional authority as dependent upon force and persuasion (Starr 1982, 13). According to Starr, the AMA was able to consolidate professional authority once Jacksonian egalitarianism gave way to the Progressive Era's embrace of scientific expertise. Seizing the zeitgeist, the AMA offered more effective ideas and carried out adroit political strategies to achieve professional power.

Starr's work rightly remains the foremost sociological account of American medical professionalization. However, while Starr's *analytical* approach of integrating both cultural and organizational factors is laudable, his *historical* analysis unfortunately reduces culture to an external context (i.e., Jacksonian democracy or Progressivism). He invokes ambiguous phrases to explain these macro-cultural mechanisms (e.g., "on the shoulders of broad historical forces [140]"). Moreover, he reproduces the logic of the truth-wins-out narrative by treating bacteriological discoveries as self-evident, ignoring the ways that its supporters worked to make them appear so. His respect for the cultural authority of science is so firm that his history suggests that once scientists got their facts straight, medicine was ineluctably transformed in ways that allowed physicians to capitalize "naturally" on the latest discovery or breakthrough. In reducing culture to an external context, Starr never turns his critical eye toward the actual production of medical knowledge.[4] For Starr, the achievement of professional legitimacy is seen as the outcome of the removal of an external cultural barrier and the elucidation of crucial facts rather than a project to create legitimacy for a particular vision of medical science.[5] This is not to pick on Starr; these analytical failings are widespread. If the poverty of the truth-wins-out logic can penetrate good sociological analyses, the oversight is systematic, the blind spot widespread.

Whither Epistemology?

The persistence of the truth-wins-out logic in the professions literature underscores the need for a more rigorous engagement with the sociology of knowledge. The existing accounts of the U.S. medical profession, both the more hagiographic versions and critical sociological analyses, fail to investigate epistemological change as an object of analysis, as a phenomenon in need of an explanation. Absent such a focus, they suffer from a basic misconception as to the nature of knowledge. Epistemological assumptions are seen as somehow timeless and outside of history. But it does not take a verdant historical imagination to see that often what is widely accepted as true in one period is dismissed as false in the next. *Standards* of truth change over time. History is strewn with the carcasses of discarded ideas once embraced as truth. Less apparent, but more significant, is that epistemological systems themselves have histories, waxing, waning, and even disappearing altogether. The life and death of ideas is not merely a matter of better or truer ideas supplanting older ones, but also of the emergence of entirely new ways of thinking.

The notion of a general linear progress of knowledge, derided by sociologists of science and challenged by historians, is undermined by the historical plurality of assumptions regarding the standard of truth against which ideas are judged. Ideas are promoted (and demoted) against the backdrop of basic assumptions about the nature of knowledge. To say an idea is accepted as true is to point out that it meets the standards of good knowledge of a particular epistemological system that develops out of social networks of thinkers (Collins 2000). Absent some basic agreement as to what constitutes legitimate knowledge, no such assessment is possible. When these assumptions change—when epistemological standards and values are jettisoned for new ones—the previous era's ideas must be translated, accommodated, or discarded. Ideas only make sense—and in turn can only be evaluated—from within an epistemological system. This is not meant to dismiss ideas as socially constructed or to relativize all knowledge claims; it is merely to contextualize truth claims, to embed ideas—and the evaluation of these ideas—within their historical-epistemological context.

But what accounts for the adoption of one epistemology over another? Changes on the level of epistemology cannot be explained away by appeals to truth and falsity. In a fundamental sense, the adoption of one epistemology over another is a matter of collective *agreement*, a typically tacit acceptance of the basic standards for evaluating knowledge claims. Put differently, when the metric for assessing truth claims changes, the ascendancy of one metric over another cannot, by nature, be determined by appeals to truth. It is not a matter of truth versus error, rational versus irrational, but rather of socially mediated choice that arises from the interaction between social actors. It was this matter of acceptance of an epistemological system based on the laboratory—and the conflict out of which it emerged—that was crucial in shaping the professionalization of U.S. medicine during the nineteenth century.

Given the centrality of this epistemological change—and the overwhelming empirical evidence in the historical record of these changes—why has epistemology fallen out of historical accounts of the professionalization of U.S. medicine? Why has the truth-wins-out narrative proven so durable? First, some of the oversight can be attributed to the paradoxical fact that even though epistemological debates are fundamental, they operate subtly. People tend not to discuss epistemological issues; rather they are the background factors that become manifest in specific debates. Doctors rarely fought over the nature of medical knowledge explicitly, but issues of the legitimacy and usefulness of certain forms of knowledge arose consistently in their specific debates over cholera. Pay too much attention to these surface knowledge debates and the epistemological subtext goes unnoticed.

Second, even when accounts of the professionalization of U.S. medicine do acknowledge epistemological changes, they commonly reverse the temporal relation between ideas and epistemology. Epistemological change is viewed as following from new discoveries. A microbe is seen; the lab is embraced. This, however, inverts temporal directionality. Before a microbe can be seen or produced in the lab, there must exist a predisposition to seeing it, an adoption of particular epistemological assumptions that would enable physicians and researchers to recognize a discovery as such. Epistemological commitments *precede* facts, not the other way around. Inverting this temporality renders invisible the role of epistemological change in the production of knowledge and the social organization of knowing.

Finally, this oversight stems from a basic lack of a historical imagination when it comes to epistemology. Commonsense, taken-for-granted notions of truth and falsity assume an ahistorical view of knowledge and truth. When one construes the standards of truth as timeless, knowledge is only relevant to the story of professionalization insofar as doctors made new medical discoveries that measured up to these standards. But this assumption is unwarranted, as standards of truth change over time. The dustbin of history is filled with previously recognized true ideas now deemed false, and there is no metaphysical warrant to assume this will not be the case in the future (Putnam 1995, 192). An epistemological system held as universal in one era gets supplanted in the next by another system that sports the same pretenses to timeless universality. From within such systems, standards seem universal, but taking the long view, we see that they are fundamentally historical.

In the past two decades, research in historical epistemology has challenged the timelessness of truth standards, temporalizing many of the attributes of knowledge and, in turn, offering a social and cultural understanding of epistemological shifts (see Biagioli 1994; Daston 1992; Daston and Galison 2010; Davidson 2001; Dear 1992; Fuller 2002; Ginzburg 1980, 1992; Jonsen and Toulmin 1988; Porter 1988; Poovey 1998; Schweber 2006; Shapin 1994; Shapin and Schaffer 1985; Toulmin 1992).[6] Historical epistemology starts from the premise that "basic epistemological categories such as cause, explanation, and objectivity are historically variable and can be studied in the same way as other types of scientific claims" (Schweber 2006, 229). Concepts like Foucault's notion of epistemes (2002), Ian Hacking's styles of reasoning

(1985), and even Kuhn's paradigms (1996)[7] serve to highlight the changing standards of truth and falsity. Epistemological units like the "fact" typically perceived as universal are shown to have a history (Poovey 1998).

In this book, I elevate epistemology to the center of my analysis, building my analysis of the professionalization of medicine in the United States upon the basic recognition of the historical nature of epistemologies. My analytical approach was born from an empirical observation made early on in my archival research—namely, that the debates between allopathic medicine and alternative medical movements during the nineteenth century were fundamentally and persistently waged on the level of epistemological claims. No matter the manifest issue, there was always a latent epistemological controversy at work. The professional politics of nineteenth-century American medicine were subsumed into an epistemic struggle that had its own rules and dynamics.

In doing so, I address a sociological puzzle: if epistemological standards change, how does this happen? A little historical imagination shows that the common answer to this question—the appeal to scientific discoveries—does not suffice. Given that epistemological standards must be in place *before* an idea is embraced, the whole notion of self-evidently true discoveries is exposed as an impossibility. Outside of historically emergent epistemological systems, ideas have no authority, no truth. The dissemination and acceptance of ideas as discoveries—as well as the professional authority that accompanies ownership of such ideas—requires an explanation *beyond the ideas themselves*. Nor can it be explained by appealing to the cultural authority of science, at least in this case. Science was not yet privileged in the nineteenth century. Indeed, there was widespread hostility toward science during the Jacksonian era (Hofstadter 1963). To consolidate professional power under the mantle of scientific medicine, reformers not only had to promote specific knowledge claims; they also had to construct the promise of science as the foremost way of knowing. The question becomes, how was this accomplished?

FROM IDEAS TO EPISTEMOLOGY

The professionalization of U.S. medicine is best understood as revolving around fundamental debates over who has the authority to speak truth and on what grounds. To develop an analysis attuned to epistemology, I marry insights from the sociology of the professions and the sociology of scientific knowledge (SSK). It is a curiosity that these two fields, so obviously related, have had little interaction over the last few decades. And while we may be able to chalk up this lack of dialogue to the idiosyncratic development of these subfields, it has created confusion about the nature of professional power and misconstrues the role of ideas in the achievement of professional authority. I seek to integrate insights from both these traditions, as they provide important concepts and points of departure to explain the rise of bacteriology and of professional authority. I focus on the important practices, or *work*, needed to bring about the new epistemological commitments integral to the acceptance of the bacteriological definition of disease and, in turn, professional power for the AMA. In the process, I offer a conceptual framework to understand the politics of epistemological change generally.

While previous sociological accounts of the professionalization of U.S. medicine have paid scant attention to the role of knowledge disputes in the emergence of the medical profession, this is not to suggest that the sociology of professions is bereft of tools for the examination of the role of knowledge in professional politics. In proposing an ecological model of professions, Andrew Abbott (1988) has sought to reorient the sociology of professions toward knowledge disputes, recognizing the overwhelming importance of interprofessional struggle in the development and decline of professions. Professions exist in an interdependent system, or "ecology" (Abbott 2005), of competitors that vie for control over certain "jurisdictions" of work. Central to these "jurisdictional disputes" are conflicts over knowledge. Knowledge is the "currency of competition" among professions, as changes to—and developments in—abstract knowledge shape a profession's jurisdiction and the system of professions generally (Abbott 1988, 102). Abbott's focus on the competition between actors reintroduces contingency and particularity into the study of professionalization. Every profession has its own unique history, its own unique struggles, and, in turn, assumes its own unique organizational form and place within the system of professions.

Because the definition of reality and the production of knowledge are central to the ability of a profession to gain control over a jurisdiction, exploring developments in a profession's knowledge base is crucial in accounting for professionalization. And while Abbott does not address epistemology directly,[8] it is easy to see how his model, focused as it is on the dynamic struggle between collective actors over knowledge, could be extended to accommodate epistemological debates. Yet, while Abbott developed his ecological model with great historical sensibility, it works best for periods in which a system of professions already exists.[9] The situation for nineteenth-century physicians, aspiring to professional authority, was much different. Not only were they struggling over professional plums; they were struggling to justify professions as a legitimate organizational form during a period when such an idea was highly suspect. In the preprofessional world, knowledge struggles revolved less around professional jurisdictions and more around basic concerns of how to organize knowledge.

Although Abbott's ecological model has become central in the professions literature—it is widely cited, its terminology extensively employed—its analytical potential has not been fully realized. No one has attempted to revisit the history of the American medical profession through a similar analytical lens. And while Abbott's heavy focus on knowledge would suggest a close working relationship between the sociology of the professions and the sociology of knowledge, there has been almost no dialogue across these subfields. Lamentably, Abbott's call for a deeper engagement in knowledge struggles—echoed in Eliot Freidson's (2001, 27) argument that an adequate sociology of work and professions "must also be a sociology of knowledge"—has gone unheeded.

SSK has much to offer the study of professionalization, especially on how actors struggle to win recognition for their ideas. Scientific practice, and by extension scientific knowledge, is shaped by the local context, cultures, and contingencies (see Collins 1992; Knorr-Cetina 1999; Kohler 2002; Latour and Woolgar 1986; Pickering 1984). Extra-scientific factors operate in the production and validation of science both *within* the laboratory, where facts are not "discovered" but produced, and outside it, in public disputes over science. Of particular relevance for the purposes of this analysis is research in what Thomas Gieryn (1999, ix) calls the "sociology of science downstream" that attends to the interpretation and consumption of science in the public sphere, "downstream" from the laboratory. To win the status of scientific truth in the public realm, scientists engage in advocacy work to capture authority and legitimacy for ideas through practices like boundary work (Gieryn 1999), network formation (Latour 1987, 1988), appeals to cultural norms of trust (Shapin 1994), the deployment of formal empiricist discourse (Gilbert and Mulkay 1984), and specific performances of expertise (Hilgartner 2000). Scientific truth is therefore not simply a status produced internally in the lab; it is created out of the dynamic struggles between actors in the public arena over how to interpret particular ideas.

While SSK deals with scientific knowledge struggles extensively, it offers less guidance in theorizing issues of epistemology.[10] This is not to suggest that SSK has been silent on epistemology. In an important sense, the entire subfield can be viewed as a direct epistemological challenge to science by showing how science does not develop by the application of procedures ratified by the type of epistemological norms that philosophers take seriously. And insofar as SSK examines struggles over science, it addresses issues of epistemology—how people achieve scientific knowledge and how disputes over scientific interpretation get adjudicated. Still, SSK rarely takes up epistemology and epistemological struggle directly as an object of analysis.[11] This oversight results from its focus on *scientific* forms of knowledge. Though contested and continuously defined (Gieryn 1999), claims to science share a core set of epistemological assumptions (e.g., a commitment to the scientific method, a notion of objectivity, the elevated status of facts, etc.) that remain unquestioned in scientific practice and even scientific disputes. Although there is some heterogeneity in how science is framed and understood (Knorr-Cetina, 1999), the underlying ideals of science constrict these divergences, and shared valuation of these ideals remains. Put differently, much of the literature of scientific disputes revolves around whether what one observes can be considered legitimate scientific evidence; epistemological disputes involve something deeper, more fundamental, namely *what evidence is*. The theoretical insights gleaned from analyses of empirical cases in which the core epistemological assumptions of science are taken for granted are limited for examining cases in which epistemological assumptions are contested, like debates over nineteenth-century medicine.

Epistemic Contests

Epistemological struggle was decisive in shaping the process of American medical professionalization. Therefore, I embed my analysis in epistemological change, investigating the practices by which actors advocated for—and struggled over—competing epistemological visions, so as to obtain a full account of how professional medical politics unfolded during this period of confusion and flux in medical knowledge and to ultimately understand the exceptional result that emerged from this political/epistemological conflict.

At its most basic, epistemology explores how we justify the things that we say we know (Kurzman 1994). As Thomas Nagel (1989, 69) points out, “The central problem of epistemology is the first-person problem of what to believe and how to justify one’s belief—not the impersonal problem of whether, given my beliefs together with assumptions about their relation to what is actually the case, I can be said to have knowledge." Epistemological questions are not just academic; they touch on a core problem of social life—by what standards can individuals adjudicate true knowledge from false opinion?

Despite the real-world implications of epistemological issues, epistemology has traditionally been constrained to philosophy, which focuses on the issue of justification in a particular manner. Committed to the conceit of universal truth, philosophers of epistemology attempt to locate a universal grounding for all knowledge. Given philosophy's disciplinary goals, philosophic discussions of epistemology assume a prescriptive, normative character, addressing how knowledge *ought* to be justified (Kim 1988; Longino 2002). They operate on an abstract plane, appealing to thought experiments to score various logical points. Arguments play with reality, asking the reader to conceive of possible worlds, filled with everything from Descartes's deceiving demons to brains in vats. Such speculation is not unproductive, as it points to logical inconsistencies in commonsense understandings of knowledge. But because philosophy's "central terrain remains conceptual rather than historical or empirical" (Collins 2000, 880), its theoretical toolkit distorts our understanding of how the epistemological problems are solved *in practice*,[12] as it lacks empirical grounding, is largely normative in nature, and is biased toward an individualistic view of knowledge. I seek to consciously relocate the discussion of epistemology from the rarefied air of philosophy to sociology by identifying the on-the-ground practices through which actors achieve epistemological change.

To do so, I develop the concept of the *epistemic contest*. An epistemic contest is one in which actors, advocating competing understandings of reality and the nature of knowledge, struggle in various realms to achieve validation for their epistemological systems, or what Ian Hacking (1985) calls "styles of reasoning." Epistemic contests are a specific intellectual type of debate in the larger "politics of cognition" (Zerubavel 1999, 22). At stake are such questions as: What is the nature of truth? What constitutes legitimate knowledge? How can such knowledge be acquired? How can beliefs about reality be justified, and who can be considered a legitimate knower? While they become manifest in controversies over specific knowledge claims, epistemic contests operate on a more fundamental plane, involving debates over which *approach to knowing* represents the most promising way to truth, and in turn, demands societal investment.

In this case, allopathic physicians and alternative medical movements not only debated specific knowledge claims (e.g., what is cholera?), but indeed the very nature of medical knowledge (e.g., what constitutes a medical fact?). As such, I use this particular case to explore *how* collective actors struggle over epistemological claims in an environment bereft of shared epistemological assumptions, identifying a number of strategies and factors that shape the outcome of such a struggle. Not only do I retell the story of U.S. medical professionalization, but I also develop the sensitizing concept of the epistemic contest so that it might be used to examine cases where similar conditions hold.[13] Identifying more or less generalizable characteristics of epistemic contests, I establish a framework to attend to the distinctiveness of knowledge disputes that go to the heart of what constitutes legitimate knowing.[14]

Epistemic contests are unique among knowledge disputes or "cognitive battles" (Zerubavel 1999, 44). Their distinctiveness stems from the depth of the challenge to knowledge they involve. Rather than debating the merits of a particular claim vis-à-vis a system of agreed-upon standards, epistemic contests involve fundamental debates over the standards themselves. Their unique character is evident when compared to Gieryn's (1999) concept of the credibility contest, to which I am indebted. In his sociology of science downstream, Gieryn examines the cultural and interpretive work performed on an idea in the public arena. His examination of credibility contests focuses on boundary work by which actors discriminate science versus nonscience. In the cases Gieryn examines, there exists a preexisting shared valuation of science—a basic agreement on the epistemic value of scientific knowledge—in which advocates of ideas want to be included.[15] In other words, there is epistemic consensus; the disputes in credibility contests revolve around whether particular claims can be said to conform to these agreed-upon standards. While epistemic contests involve issues of credibility, they are broader and more basic, incorporating questions as to what is the legitimate way by which people come to know reality. Preexisting epistemological agreements are absent, and science may be only one of the many epistemological systems involved.[16] To clarify this distinction, two experts debating over the interpretation of experimental evidence is not an epistemic contest; two individuals debating the legitimacy of experimentation is. The concept of epistemic contests shifts attention to debates in which more elemental issues of knowledge are at stake, not just who can speak with scientific credibility. Indeed, as epistemic contests involve situations lacking common cultural assumptions, they are waged through diverse strategies that contain both organizational and cultural components. Therefore, in reconstructing the epistemic contest over nineteenth-century medicine, I avoid the false dichotomy between organizational and cultural practices, revealing how the strategies adopted by collective actors in epistemic contests contain elements of both, and by extension, suggest that all knowledge disputes, even those read as solely cultural, do as well.

Every epistemological system contains a working model of what constitutes legitimate knowledge, the method(s) by which such knowledge can be attained, and the general ethos and identity that legitimate knowers should possess in relation to knowledge. They outline the assumptions that guide knowledge production and adjudicate competing claims. Epistemological claims therefore can be made along a number of dimensions. Actors can aver epistemic authority on account of the *content* of their knowledge, arguing that their possession of a body of knowledge legitimates a privileged epistemic position. Or epistemic authority can be rooted in claims to possess the *method* by which knowledge is produced. Methods-based claims can represent some sort of abstract ideal, like the scientific method, or more specific forms of technical acumen, like the technical know-how to process certain forms of information. Finally, epistemic authority can be justified along the lines of the orientation or *ethos* the group assumes toward knowledge. This ethos typically underwrites claims to epistemic authority based upon objectivity, where appeals to disinterestedness, trustworthiness, and a "view from nowhere" (Nagel 1989) become common tropes by which actors claim epistemic authority. Naturally, in practice these dimensions overlap, but it can be useful to parse them analytically, as the specific way in which claims are made shapes the strategic choices of actors.

However epistemic authority is framed, the concept of the epistemic contest directs attention toward conflicts among competing epistemological systems. Such conflicts, though not commonplace, are particularly disruptive. Epistemological assumptions form the core of the taken-for-granted world. These ever-present criteria for assessing beliefs inform and determine the manner in which individuals make sense of reality. In most cases, they remain unarticulated. Individuals do not need to be able to articulate justificatory arguments in order to employ them in the pursuit of knowledge (BonJour 1978). Such standards are institutionalized in the social practices of knowing however internalized and unconscious they may become. They are rarely questioned, much less discarded. For this reason, knowledge disputes typically do not operate on the level of epistemology. But when they do, the urgency of the debates is intense.

To overcome the inertia of commonsense thinking, epistemic contests require certain conditions. Crises and major disruptions are often necessary in establishing opportunities for epistemic contests, but they are not sufficient. In the case of American medicine, cholera led to an epistemological crisis; the great influenza did not. Still, even with such exogenous crises, it takes significant *work* to destabilize epistemological assumptions—work performed by actors competing for recognition as legitimate knowers. The stress on the agency of actors in waging epistemic contests is paramount. Context matters, but it is what actors *do* that affects the trajectory of epistemic contests. It is true that the epistemic contest over medicine was enabled by certain long-term structural changes in the United States during the nineteenth century. And it is true that the arrival of cholera created a sense of panic and urgency that exposed some fundamental fissures within allopathic medicine. But neither was decisive in creating an epistemic contest. Rather, as I show, alternative medical movements seized the opportunity cholera provided to demand that allopathic physicians give an epistemological account of their authority as knowers. They forced the specific medical debates that cholera elicited onto the level of medical epistemology, thus shaking the foundation of allopathic medicine and giving birth to a nearly century-long struggle over what medical knowledge consists of. Cholera offered alternative medical movements an opportunity to assert themselves vis-à-vis allopathy. They transformed it into an epistemic contest.

Though constrained by structural factors, epistemic contests unfold according to their own internal logic, rendered contingent by the strategic back and forth between competing actors.[17] Actors engage in what I have called elsewhere "knowledge advocacy" (Whooley 2008), championing certain versions of knowledge in a struggle to achieve epistemic authority and gain recognition. These visions clash, for the knowledge advocates argue *against* alternative visions of knowledge as much as they argue *for* their own. Interactive struggle then is central to explaining the trajectory of epistemic contests and its resultant effects on knowledge production. The epistemic contest over medicine evolved according to the dynamic, give-and-take between allopathic physicians, alternative medical movements, and medical reformers, as it traversed different organizational settings and involved a wide array of actors. Medical sects did not present fully formed articulations of their epistemologies and stick to them over the course of the contest. Rather, their epistemological positions arose in relation to each other and were modified over time according to their interaction.

The stakes of epistemic contests are great. When one epistemic system is elbowed out by another, epistemic closure—the ascension and dominance of one epistemic system over others—is achieved. What is forfeited in closure is an entire way of understanding the multitudinous experiences of reality, and in turn, any future insights that alternative styles of reasoning might offer. The possible forms that knowledge can assume are restricted. While productive and necessary, epistemological systems all have their blind spots, allowing for certain types of questions and answers to arise while forbidding others. Thus, epistemic closure does not just result in the selection of certain ideas over others; it involves the selection of an entire approach to knowledge at the expense of others.

The legacy of epistemic contests is evident in the power disparities—cultural and organizational—they leave in their wake. Solutions to the problem of knowledge are solutions to the problem of order (Shapin and Schaffer 1985). In the resolution of epistemic contests, certain actors are granted trust by society, while others are denied as legitimate knowers. With victory comes enticing spoils—cultural capital, formal legitimacy, organizational resources, and institutional support. These spoils become institutionalized, as the taken-for-granted practices of these organizations—"how institutions think" (Douglas 1986)—come to embody the winning epistemology, and subsequently disseminate it through processes like isomorphism (DiMaggio and Powell 1983). As such, the power disparities in terms of who is and is not recognized as a knower become imbued with an inertial quality as the patterned thinking is transformed into common sense.

Because professional privilege is one of these spoils, epistemic contests can be implicated in professional politics. Ultimately, both professional struggles and epistemic contests are about the power to define the real, and as such commonly overlap. Insofar as professions represent a privileged economic position granted on the basis of specialized knowledge, defining the standards and nature of knowledge is crucial for professionals. This is not to say that professional struggles *necessarily* involve epistemic contests; professional debates need not involve fundamental questions over the nature of knowledge. They can revolve around a whole host of other issues (e.g., organizational control). Nor do epistemic contests necessarily spur professional struggle; they can—and do—occur outside of the system of professions. However, when professional struggles take on the form of epistemic contests, as was the case in nineteenth-century medicine, epistemology becomes decisive.

In the end, the analytical payoff of the concept of the epistemic contest comes from its emphasis on embedded struggle over time in accounting for epistemological shifts. Rather than offering teleological accounts of the development of knowledge, I offer a conflict model of knowledge production that is shaped by the strategic give-and-take between actors, strategies shaped in part by the organizational contexts in which they unfold. Unlike most histories of this medical period, which take the point of view of particular medical sects, I elevate the *interaction between actors*—what William Sewell (2005, 6) refers to as "unfolding of human action through time"—to the center of the analysis. To understand the eventual consolidation of medical authority around the bacteriological model of disease, we must understand how and why medical sects made certain choices, and adopted certain strategies, in response to the actions of their challengers. At its core the dispute over medical knowledge and epistemic authority in the nineteenth century was *relational*, emerging out of the dynamic dealings between the actors involved. This is not a history of allopathy *or* homeopathy *or* any of the other actors described herein, but rather of their contentious relationship. In elevating the role of struggle—in stressing politics *and* knowledge—I show that the eventual content of medical knowledge had its birth, not fully in an external reality, but in the legacy of intellectual debates and extra-scientific polemics in which it was embroiled. The development of medical knowledge did not involve a simple reading of an external world or the discovery of certain facts; it involved a struggle to control and define the very standards by which knowledge was to be judged. And out of such epistemological struggle was born a unique and exceptional institution—the modern U.S. medical profession.

WHY CHOLERA?

This book recounts how regular physicians won the epistemic contest over medicine and, in turn, were able to consolidate professional authority, lost after the 1832 cholera epidemic, under the bacteriological paradigm in the early twentieth century. First, it explores the origins and development of the epistemic contest. How did alternative medical movements create an epistemic contest that translated into successful legislative campaigns between 1830 and 1890? Second, it describes the process by which regulars, through the AMA, were able to defeat all epistemic challengers to create an exceptionally powerful profession. How did allopathic physicians consolidate their professional authority around the bacteriological paradigm so as to legitimate the development of an institutional structure that excluded other medical sects? How did such epistemic closure shape the modern medical profession?

To consider the establishment of an epistemology in general, to follow it through its various arenas, to note the number of actors involved in debunking or championing it, and to trace its trajectory over eighty years would be too large a task to furnish the type of close analysis necessary to understand the struggle for epistemic and professional legitimacy in all its subtlety and complexity. Fortunately, such a herculean task is unnecessary. Epistemic contests become manifest in specific debates. And not all medical debates are created equal. I have re-created the history of the epistemic contest through the case study of cholera during the period between 1832 and 1912—a period that witnessed four cholera epidemics in the United States.[18] The data for this book has been culled from a wide array of primary source documents (e.g., medical journals, speeches, pamphlets, private papers, sociology meeting minutes, etc.). Using this extensive body of archival data, I painstakingly reconstruct the epistemic contest through a content analysis of both the specific claims of cholera being made as well as the epistemological assumptions underlying these claims. In doing so, I recount the muddled history of the epistemic contest over medicine in general, beginning with the successful challenge of alternative medical movements in the mid-1800s and ending with the allopathic achievement of epistemic closure under the epistemology of the laboratory at the turn of the century. Put differently, I provide a historical/sociological analysis of the politics of nineteenth-century medicine with cholera as its focus.

The choice of cholera results from its tremendous historical importance—it was *the* medical issue of its day—as well as from some pragmatic methodological concerns. Historically, cholera was a major cause of the epistemic contest; analytically it offers a mirror to the more general problems animating nineteenth-century medicine in the United States. And while cholera was not the only issue involved in the epistemic contest,[19] as "the classic epidemic disease of the nineteenth century" (Rosenberg 1987b, 1), it was arguably the most important in both the reach of its influence and the persistence of its threat. As an exogenous shock to the medical profession, cholera disrupted the traditional workings of U.S. medicine and presented a host of problems, epistemological and otherwise, for the medical profession.

Thus, while the epistemic contest became manifest over a number of different issues (e.g., the therapeutic value of bloodletting) and disease definitions (e.g., tuberculosis, yellow fever), cholera as a case has a number of benefits over other possible issues. First, epidemics are useful "sampling devices" as they bring to the fore many social, economic, political, and intellectual issues that are less visible during more tranquil periods (Rosenberg 1966). There exists a rich tradition of using cholera as a sampling device to study larger social phenomena. For example, Charles Rosenberg's seminal study *The Cholera Years* (1987b) explores the secularization of American society by showing how the understanding of cholera evolved from the scourge of the sinful to a consequence of remedial faults in sanitation.[20] Additionally, researchers have used cholera as a lens to examine a number of issues, including the sociopolitical history of nineteenth-century Hamburg (Evans 2005); the evolution of the bourgeoisie in postrevolutionary France (Kudlick 1996); political conflict in Lower Canada (Bilson 1980); the vagaries of British class politics (Durey 1979); the cultural norms of Victorian England (Gilbert 2008); resistance to tsarist policies in Russia (Friedan 1977); and the role of medicine in reinforcing American nativist policies and politics (Markel 1997). I use cholera as a sampling device to explore the intellectual crises within medicine in the early nineteenth century, taking advantage of the fact that cholera remained a pertinent issue over the entire period of professionalization that I seek to understand. While certainly cholera loomed larger during the initial period under study (1832 to 1866), it remained a persistent puzzle for medicine into the twentieth century. And it forced allopathic physicians to reassess their epistemology by offering opportunities for the elaboration of alternative visions for medical epistemology. When it came to cholera, the intellectual and professional stakes were high.

While most of the historical research on cholera focuses on other national contexts, historians have long recognized cholera's significance on American medicine. For nineteenth-century physicians, there was no medical problem more vexing and more significant than cholera. Cholera's importance is alluded to (if not expounded upon) in the historical scholarship on the disease. The professional status of allopathic physicians depended greatly on achieving an understanding of cholera (Rosenberg 1987b). Likewise, the consensus among historians of alternative medical movements is that the failure of regular physicians to stem the tide of the 1832 epidemic was used by alternative movements to challenge allopathy (Berman and Flannery 2001; Coulter 1973; Haller 2000; Kaufman 1988; Whorton 1982). And many of the doctors who would eventually lead the charge in reforming American medicine cut their teeth on the study and treatment of cholera, especially in Paris in 1832 (Warner 1998).

Second, the physical reality of cholera caused it to dominate the public imagination far beyond its actual mortality rates. Few diseases can match cholera in the speed and intensity with which it kills. Symptoms and biology play a role in the panic elicited by an epidemic as they offer an underlying reality to historical experience (Humphreys 2002). The physical event of cholera was (and is) dramatic and terrifying; it was (and is) a shocking disease. According to the current understanding of cholera, the cholera toxin paralyzes the intestines, causing the intestinal cells to rapidly secrete water and electrolytes. The body purges copious amounts of rice water stool—up to 10 percent of a person's body weight within hours. Severe dehydration sets in quickly, causing intense muscle cramps, sunken eyes, and a bluish tint to the tongue, lips, and other extremities. If left untreated, 70 percent of its victims die. An individual in his or her prime can be dead within ten hours. Therefore, while the mortality of cholera may not have been as severe as some other endemic and epidemic diseases, its biological nature caused it to loom large in the nineteenth-century medical imagination.

Finally, as suggested above, the way in which cholera became redefined as a medical problem confounds the conventional narrative of the rise of the bacteriological model of disease. The eventual consolidation of medical authority around the bacteriological model of cholera does not fit the straightforward truth-wins-out narrative for a number of reasons:

- First, Koch's famous discovery of cholera was beset with inconsistencies. It did not even satisfy his own postulates. While certainly other disease entities that were widely accepted also fell short of Koch's postulates, the failure of cholera to meet these standards shows that it was not a natural fit for the bacteriological model.
- Nor is it clear that the bacteriological definition was more effective in treating cholera. Unlike other diseases such as diphtheria and rabies, in which a bacteriological model led to effective vaccines, a cholera vaccine was never popular. Treatment of cholera is fairly simple—a patient is given a large amount of saline injections to rehydrate. This therapy targets the

symptoms of cholera—copious diarrhea and dehydration—and does not depend on an understanding of its etiology.

- In terms of prevention, the miasmatic theory of cholera—the notion that disease is caused by miasma (pollution), or noxious "bad air" in the atmosphere—was more effective in eradicating cholera as it led to many of the sanitary reforms responsible for the disease's demise in the United States (Duffy 1990; Rosenberg 1987b; Tesh 1988). Even the self-proclaimed champion of the "new germ theory of disease," Paul Ewald (2002, 77), acknowledges that the water treatments, adopted under the miasmatic theory, were (and are) the most useful approach to cholera. Recent calls to reintroduce environmental factors into the study of cholera represent a shift away from the reductionism of the germ theory toward the "biocomplexity" of the miasmatic theory (Colwell 2002; Colwell and Huq 2001).
- Finally, recent evidence shows that the bacteriological understanding of cholera is not as clear as once depicted (Hamlin 2009), questioning the argument that the bacteriological model is the objectively "right" one. In analyzing cholera's killer instinct, Waldor et al. (2003) identified two components of cholera's attack that facilitate its rapid spread—(1) the TCP pilus in cholera vibrio that allows it to replicate rapidly and (2) the cholera toxin that triggers rapid dehydration. The gene for the cholera toxin is actually supplied by an outside source—a virus called CTX phage. Without this gene, cholera does not know how to be a pathogen (Johnson 2006). This classic case of coevolutionary development of two different organisms raises the question, is cholera caused primarily by the bacterial microorganism cholera vibrio, or by the virus called CTX phage? As researchers discard the germ-in-the-laboratory model of cholera for a more environmental approach (Hamlin 2009), does this simple etiological question even make sense?

While certainly not enough to dismiss the truth of the bacteriological model of cholera, these issues introduce more uncertainty and messiness than is common and, in turn, undermine the straightforward model of dissemination and acceptance of past research. To understand the emergence and acceptance of the bacteriological definition of cholera, we cannot fall back on an argument of therapeutic or preventative efficacy. Nor can we account for this model by pointing only to bacteriological success in explaining other diseases. Like any paradigm, the bacteriological model fits better for some cases than others; to understand why it became the universal model for *all* diseases requires that we investigate those diseases, like cholera, for which it was problematic.[21]

EXPERT KNOWLEDGE IN DEMOCRATIC CULTURES

The epistemic contest over cholera witnessed four cholera epidemics, the emergence of a number of alternative medical movements, rancorous legal debates, government resistance to professional claims on democratic grounds, untold etiological theories of cholera, and sadly, rampant death at the hands of enthusiastic, but unknowledgeable physicians. By tracing the intellectual debates over cholera through various institutions (e.g., state legislatures, boards of health, professional societies, etc.) and among a diverse set of actors (e.g., allopathic physicians, homeopaths, public health reformers, etc.), I demonstrate how regular physicians, through the AMA, were able to overcome all of these challenges to create a powerful profession unfettered to the whims of the state and the vagaries of democratic decision-making.

Each chapter is organized around a pivotal moment in the history of cholera, exploring issues pertaining to epistemological politics so as to develop the concept of the epistemic contest. The first two chapters describe the initial confusion surrounding cholera, the decline in authority for allopathic medicine that resulted, and the limited allopathic response to this professional/epistemic crisis. Chapter 1 focuses on the effective campaigns of alternative medical sects to transform the perceived allopathic failure during the first cholera epidemic into an epistemic contest that eventually led to the wholesale repeal of licensing laws. Thomsonism, an egalitarian, anti-intellectual grassroots medical movement, and homeopathy, an elite urbane sect that sought to claim the mantle of science, offered more democratic epistemological visions for medicine that contrasted with the elitist, obfuscating epistemology of rationalism. They compelled regulars to provide an epistemological justification for their professional privileges in state legislatures. Drawing on theories of rhetoric, I show how the democratized epistemologies of alternative medical sects resonated, rhetorically and epistemologically, with the state legislatures influenced by Jacksonian ideals. Licensing laws were universally repealed in the 1840s; the medical market was deregulated; and an epistemic contest was born.

Chapter 2 describes the allopathic response to the democratic challenges of alternative medical sects, particularly homeopathy. After the 1848 epidemic, allopathic reformers redefined the identity of regulars, embracing a radical empiricism inspired by the Paris School of medicine. While this shift ostensibly allowed allopaths to claim some democratic bona fides, the selective manner in which they adopted the Paris School led to intellectual fragmentation. Eschewing the search for general laws in medicine (Warner 1998), allopathic reformers lacked standards to adjudicate competing knowledge claims. To solve this "problem of adjudication" they adopted an organizational strategy, establishing the AMA and substituting the criterion of membership for epistemological standards in order to deem homeopathy as quackery. Still, the exclusionary politics of the AMA failed to sway legislatures, which remained committed to the idea that open debate would lead to the best medical knowledge. This chapter reveals that epistemic contests are not waged by cultural/epistemological means only; organizational strategies can be usefully analyzed as epistemic practices as well.

Chapter 3 discusses a key event in the history of cholera—the establishment of the Metropolitan Board of Health of New York City and the rise of public health more generally prior to the 1866 epidemic. United around a common understanding of cholera as a miasma, an eclectic group of actors, which included sanitary-minded allopathic physicians, homeopaths, social reformers, and sanitarians, came together to prevent cholera by cleaning up the environment. This was accomplished to great result, and the board of health was widely credited with having prevented another cholera epidemic in New York City. As public health grew in popularity, allopathic physicians sought to transform sanitary success into justification for their professional recognition. Once again, the legislatures refused to recognize these claims, as sanitarians framed them as contrary to the apolitical nature of the public health enterprise. Public health remained an eclectic movement rather than an allopathic-dominated one. This chapter explores the multiple ways in which claims to epistemic authority can be made, noting that *how* actors choose to make these claims has ramifications for their professional goals.

The final two empirical chapters explore the consolidation of allopathic professional authority through epistemic closure. Chapter 4 describes the ways in which American physicians interpreted the "discovery" of the comma bacillus of Robert Koch in 1884. Drawing on an "attributional model" of discoveries, this chapter explores the role of "discoveries" in epistemic contests, showing how the project to configure Koch's research into a discovery involved both cultural and organizational dimensions. Both homeopaths and allopathic physicians initially staked a claim to Koch's research, attempting to frame this research into a discovery that justified their respective systems of medicine through different discovery narratives. I show how allopaths offered a more effective discovery narrative, which facilitated the construction of a network linked to German science and allowed them to claim Koch as their own to the detriment of homeopathy.

The final empirical chapter discusses the consolidation of allopathic professional authority and elimination of sectarian threats. Allopathic reformers sought epistemic closure through an epistemology of the laboratory, based on the germ theory of disease[22] and the laboratory sciences imported from Germany (Bonner 1963). This approach redefined cholera as a microorganism, identified in the lab through the microscope, and treatable through vaccines, antitoxins, and inoculations. Despite this reframing, laboratory analysis was routinely ignored during the 1892 cholera scare. Cognizant of government skepticism and the limitations of achieving professional recognition through public health, the AMA adopted a conscious program to circumvent government institutions by aligning itself with private philanthropies. Reformers found allies among industrial philanthropists who were beginning to integrate the laboratory into their businesses and eventually convinced them to fund their program of scientific medicine. Using these philanthropic resources, allopathic reformers were able to make the laboratory the "obligatory passage point" (Latour 1987, 132) for all medical knowledge, to create an organizational infrastructure around the lab under their control and purified of homeopathic influence, and achieve the standardization of medical education along bacteriological lines. Epistemic closure was achieved by allopaths without having to debate the merits of their system in the democratic public institutions where they had been continuously defeated.

While it is difficult to reconstruct the motivations of actors long dead, especially since the issue of motivation is best approached on an individual case-by-case basis, I'd be remiss not to say something about how I conceive of the actors in this book. These physicians should not be reduced to cynical political operatives. Nor should they be romanticized as disinterested seekers of truth. Between these two extremes lies a more balanced depiction of actors with multiple (often conflicting) motivations. The shifting commitments to particular epistemological systems by nineteenth-century physicians were driven both by a desire to solve intellectual problems and a desire to gain a strategic advantage in the epistemic contest. Physicians, whatever their sectarian allegiance, strove to make sense of cholera while also gaining recognition and power. It is this messy combination of noble truth-seeking and base politicking that makes epistemic contests so compelling.

The particularities, and peculiarities, of the professionalization of U.S. medicine facilitated the rise of an exceptional medical system, unusual in the developed world. Although a comparative analysis of the organization of medicine in different countries is beyond the scope of this project, suffice it to say that the U.S. medical system is widely viewed as an odd duck. The twin pillars of this exceptionalism—its embrace of private interests and its wholesale adoption of a scientific vision of medicine—originated from the particular trajectory of the epistemic contest that resulted in a profession highly suspicious of government involvement and democratic oversight. In the end, the key emergent theme in the history of U.S. medical politics is the tension between professionalization and democracy, which I discuss in the conclusion. The former stresses the recognition of a protected, privileged group of experts, in which the production of knowledge is mystified and insulated from public oversight. The latter stresses transparency and participation. The animating issue underlying the epistemic contest over medicine was the question of the place of expert knowledge in a democratic society. Indeed, without doing too much injustice to the nuance of the analysis, the entire epistemic contest could be read as an account of the persistent tensions between democratized epistemologies and the exclusive epistemological system proffered by allopathy. Skirting the public institutions of the state, allopathic physicians overcame democratic debate by avoiding it, persuading a small group of elite philanthropists to bankroll their professional project. The success of this "strategy of nondialogue" (Biagioli 1994, 216) was not lost on the AMA; it became its default strategy in subsequent debates over health care and public health in the first half of the twentieth century, and an ingrained part of its professional culture.

Focusing on the epistemic contest over medicine as determinative in the professionalization of U.S. medicine opens an analytical space for understanding the nature of the ascendancy of the bacteriological model and the triumph of allopathic medicine over other alternative medical sects. The history of this epistemic contest yields a surprising, and disconcerting, finding, namely that the genesis of the U.S. medical system involved a repudiation of democratic principles. It also serves as an example of what the sociology of epistemologies (Abend 2006, 3) can achieve, what an empirical focus on epistemology can tell us about the power/knowledge nexus (Foucault 1980). By taking a quixotic journey into the history of American cholera, we gain insight not only into the strange world of professionalization of U.S. medicine but also into the general politics of knowledge and the everyday practices of epistemology in democratic cultures.

1
CHOLERIC CONFUSION

WHEN CHOLERA FIRST ATTACKED EUROPE IN 1831, PHYSI- cians were caught so unprepared that they struggled to even name the new malady, much less prevent its spread.[1] Among the names suggested were "cholera asphyxia," "spasmodic cholera," "malignant cholera," "bilious cholera," "convulsive nerve cholera," "hyperanthraxis," and the particularly poetic "blue vomit" (Longmate 1966, 66). Eventually, the disease was anointed "cholera," a curiously misleading choice, given the amount of baggage the term bore. Under the centuries-old Hippocratic system, cholera referred to an excess of yellow bile (Hamlin 2009, 19). Over time this humoral definition morphed into a more generic stand-in for milder diarrheal diseases. Now, in the panicked days of the first pandemic, cholera underwent another definitional transformation—from "a transitory state of one's constitution" to "a relentless and deadly invader" (Hamlin 2009, 20). This hasty christening caused much confusion among physicians and officials. The victimized poor, on the other hand, suffered no such appellative confusion; to them the new disease was known simply as "the pestilence."

Whatever its name, the new disease killed in a dramatic fashion. Doctors marveled at the speed at which cholera claimed those in their prime (Rosenberg 1987b). According to Dr. M. Magendie (1832, 6), of Sunderland, cholera "cadaverizes *in an instant the person whom it attacks*." Victims purged an abundant amount of "rice water" diarrhea. A "loose and relaxed state of the bowels" was attended "by frequent loose or watery discharges" (Atkins 1832, 65) with up to 10 percent of a person's weight lost within hours. But cholera's most macabre symptoms were the "cholera voice" and the surreal color of its victims. Patients took on an eerie bluish pallor just before dying, a ghastly visage of impending death. And the blue victims emitted strange sounds. One haunted doctor reported in the *Boston Medical and Surgical Journal* (*BMSJ*), "In the most deadly form of cholera there is a tone of voice, a wail, which once heard, can never be mistaken; by him, upon whose ear it has fallen in the accents of anguish, it can never be forgotten" ("Cholera Voice," 1832, 148). More than the number dead, it was the nature of cholera that caused it to loom large in the popular imagination (Humphreys 2002).

Cholera plunged Europe into turmoil. Hungary and France each lost over one hundred thousand people to the disease. Cholera claimed another fifty-five thousand in England. These raw mortality counts only hinted at the horrors on the ground. Europe's inadequate infrastructure of charitable organizations and government institutions was overwhelmed. Churches were converted into makeshift hospitals, while their cemeteries swelled. A report from Paris described the deteriorating scene: "The deaths are so numerous every day that hearses have become altogether inadequate to the purposes for which they are ordinarily used, and the dead are carried to their burial places in large wagons" (*BMSJ* 1832b, 254). Cholera killed its victims quicker than communities could bury them (Grob 2002, 108). Wherever it touched, cholera produced a type of "epidemic psychology" (Strong 1990) of suspicion, fear, and stigmatization. The feeble actions adopted by European governments heightened tensions, often resulting in riots fueled by rumors of physician-led conspiracies against the poor (Briggs 1961; Burrell and Gill 2005; Durey 1979; Morris 1976). Faced with the breakdown of the social fabric, many looked to the heavens, conjuring up supernatural explanations for cholera. Seizing this opportunity, shrewd religious leaders used the scourge to admonish their flocks for their moral laxity.

Anticipating cholera jumping the Atlantic, American physicians scoured reports from Europe for any useful information on the disease. The editors at the *Boston Medical and Surgical Journal* (1831a, 5) instructed every American physician "to watch with eagle eye the progress of this dreadful malady, and to treasure up in his mind every incident in its history which may aid in forming philosophical views with regard to its treatment." But the profuse reporting offered scant medical intelligence (Hamlin 2009, 110). Measured analyses were difficult given the circumstances. Helpless in the face of "King Cholera," European doctors could not agree on the most basic details of the disease. Was it a new disease or a more virulent form of an old one? What was its cause? Did it prey on the weak and immoral or kill indiscriminately? Which treatments were most effective? As doctors impotently mulled these questions, deaths accumulated. Hope dwindled. Having tried "every means sanctioned by recorded experience," a London physician voiced the futility felt by many European doctors: "To our patient, laboring under a violent and advanced attack of Spasmodic Cholera, no solid expectation of recovery could be extended" (*BMSJ* 1831a, 8). Eventually, American physicians gave up on gaining any insight from abroad. An anonymous letter to the *BMSJ* (1832a, 189–190) summed up the situation:

> We have nothing, therefore, to learn from the practice of the most distinguished physicians in Europe, except to notice their errors, and to avoid the rocks and shoals upon which they have made shipwreck. Let us turn these scenes of horror to the writers of our own country. . . . Seeing the utter failure of the European physicians, in their treatment of the present epidemic, it behooves our practitioners to make themselves masters of all principal writers of their own country, who have been familiar with cold, sinking febrile disease.

While few doctors clung to the hope that cholera would not reach the United States or that the country's salubrious environment would limit its spread, most resigned themselves to the fact that they would get the chance to see the disease for themselves, as the Atlantic Ocean was no longer an insurmountable barrier for Europe's problems given advances in sea travel.

On June 26, 1832, the inevitable occurred. Cholera arrived in New York City, by way of Canada. An Irish immigrant named Fitzgerald came down with a strange intestinal illness. Dr. Cameron, a New York physician,

> found him [Fitzgerald] violently affected with vomiting, purging, and most convulsive spasms; the features sunken and the eyes staring; the pulse insensible at the wrist, and the surface cold, and covered with clammy sweat; the countenance black and terrific; tongue of a dark purple during spasms, becoming opalescent as the spasmodic action abated; the fluid rejected was watery, consisting probably of the liquids he was permitted to drink; his dejection resembled rice water, of the consistence of cream. (*BMSJ* 1832d, 354)

Undoubtedly a case of cholera. Fitzgerald recovered, but his wife and children contracted the disease and died. Cholera quickly spread through the poorest districts of the city, with the infamous Five Points neighborhood—a place that Charles Dickens (2000,101) described as encompassing "all that is loathsome, drooping, and decayed"—suffering the brunt of the attack (Grob 2002, 105). Initially, city officials debated whether or not to announce cholera's arrival. The stakes in such a decision were high. Officials, worried about the economic, political, and social ramifications of such an announcement, dawdled (Duffy 1968). In response, the city's medical society accused the board of health of being unconscionably slow in alerting the public, resulting in unpreparedness and unnecessary death. This spat fueled panic among the public, as it appeared that officials could not agree on even the most basic of issues—whether or not cholera had reached the city. No one seemed to know what they were talking about. Dr. David Meredith Reese (1833, 3) recalled,

> The great ignorance of the unprofessional portion of our population on the subject [cholera] was obviously the prolific source of much imprudence, and threw the timid into a consternation and terror which prevented the adoption of any uniform and rational mode of prevention; while, at the time, the vague as well as contradictory opinions which have found their way into the public press, upon the subject of the causes, prevention, and cure of Cholera, have been very far from inspiring confidence in the members of our profession; and in such perilous times, this confidence is more than ever necessary and important.

Cooler heads were failing to prevail.

While officials argued, many abandoned the city to cholera's chaos. John Pintard, a successful merchant, respected philanthropist and former secretary of the New York Chamber of Commerce, documented the deteriorating scene in letters to his daughter. Early in the epidemic, on July 3, Pintard expressed skepticism toward the "unnecessarily alarmed" doctors, opining that "at best we are likely to have a sickly season but we are not timid & shall stand our ground" (Pintard 1832, 66). Steeled by his military pedigree and strong Huguenot faith, Pintard refused to flee the city, unlike many of his fellow New Yorkers. Independence Day celebrations were canceled. Only churches remained open for mourners to pray for their victims, thus transforming the celebratory holiday into a somber occasion of fasting and prayer. By July 8, Pintard (1832, 69) observed, "The city is much deserted & the panic prevails." Still Pintard stood fast, comforted in the belief that cholera only attacked "intemperate dissolute & filthy people" (Pintard 1832, 72). By the end of the month, however, even this false hope was dashed, as he reported the deaths of a friendly neighbor, a "hard-working" mechanic, and three physicians. Faced with the magnitude of the epidemic, Pintard, like many, turned to religion. Although President Andrew Jackson refused

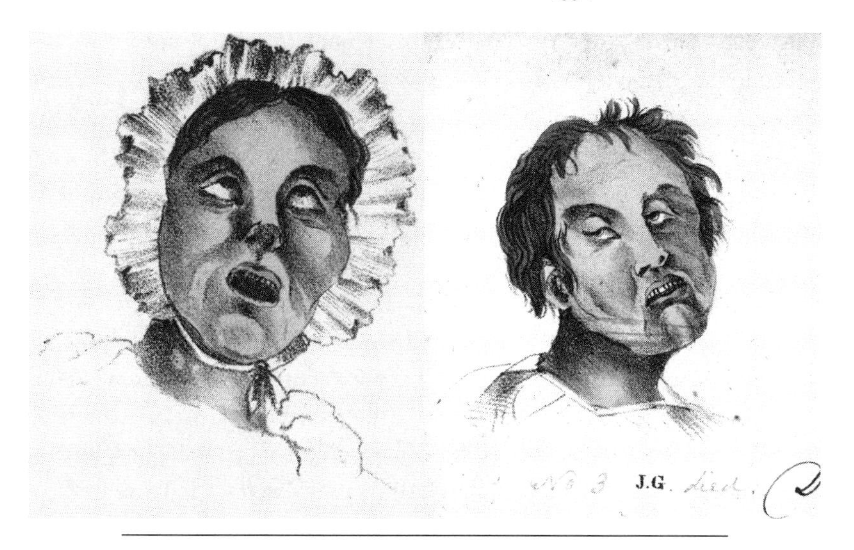

Faces of cholera, from Horatio Bartley, *Illustrations of Cholera Asphyxia in Its Different Stages, Selected from Cases Treated at the Cholera Hospital, Rivington Street* (New York: S. H. Jackson, 1832). Collection of the New York Historical Society.

to proclaim a national day of prayer and fasting, state governments did, and Pintard abided by New Jersey's July 26 day of fasting.

While Pintard captured the panic among the living, an apothecary named Horatio Bartley (1832) memorialized the dead, sketching haunting images of cholera victims from the Rivington Street Hospital. His sketches show blue-hued, skeletal faces, writhing in pain. The victims' sunken eyes betray no awareness, revealing another of cholera's more unnerving symptoms—"the entire loss of all consent, sympathy, or catenation . . . between the brain and the nervous system, and the heart and the sanguiferous system" (*BMSJ* 1833a, 271). Victims were literally severed from their corporeality, assuming a ghostly countenance.

As summer went on, the blue visages of cholera became a common sight throughout the United States. By the end of the epidemic in New York City, where cholera deaths reached their peak on July 19, 3,515 people had died out of a population of 250,000. With riverboats its chief mode of transportation (Chambers 1938), cholera spread south, first toward Philadelphia and reaching the South in late August. By late September cholera had "extended as far south as Edenton, North Carolina, and westward to St. Louis" (*BMSJ* 1832e, 253), with New Orleans suffering the brunt of the outbreak. It also traveled westward via the Erie Canal and the Cumberland Road (Grob 2002, 105). Local attempts to pull together ad hoc medical committees to mount a defense were unsuccessful. Towns employed a variety of measures to prevent the epidemic to no avail, constructing roadblocks, imposing quarantines, even shooting cannons into the air in an effort to alter the poisonous atmosphere. Wheeling, West Virginia, undertook one of the more unusual plans:

> To test the virtue of coal smoke and heat in staying the epidemic, cart loads of coal were deposited at intervals of fifty yards along each side of the principal streets and fired; the volumes of dense black smoke enshrouding the town—deserted streets, except by the frequent funeral train—sorrow and alarm depicted on every face, formed a scene more easily imagined than described; its impressions are still very vivid in my mind. (Hildreth 1868, 228)

Of the largest U.S. cities, only Boston and Charleston were spared.

Yet, as quickly as it came, cholera went. On August 15, the board of health in New York started closing down its cholera hospitals and two weeks later, it disbanded the Medical Council that had been established to combat cholera. By October, the disease had all but disappeared, and while sporadic cases were reported in 1833 and 1834, it would be fifteen years until the next epidemic. For those who lived through the epidemic the experience would linger. Pintard (1832, 92) admitted, "I shall never forget the solemn impressions of the late dreadful month of July, when the face of heaven appeared to be obscured with a somber shroud of pestilence and death." Still, life returned to normalcy. However, for physicians, the cholera epidemic would have ramifications far beyond painful memories, as it ushered in a long period of crisis for American medicine. Public confidence in regular medicine waned (Berman and Flannery 2001; Whorton 1982), as doctors were blamed for their inability to combat cholera and accused of fleeing in cowardice during their patients' time of greatest need. American doctors' opportunity to observe the workings of cholera in their own country did little to demystify the disease or assuage panic. To the contrary, it raised fundamental questions about the adequacy of medical knowledge. An anonymous letter to the *BMSJ* (1833b, 314) voiced the dismay of many physicians:

> Numerous are the pamphlets and compilations already before the public, detailing the extensive ravages of this destroyer of mankind; and yet how

> little, in view of all that has been written, worthy of retention! What has hitherto been laid down in regard to the proper mode of treating epidemic cholera? To what source shall we direct the inquiring student for the gratification of his laudable curiosity, and the establishment of his views upon the best method of combating this disease? Upon this branch of the subject, previous accounts are irregular and contradictory. The little that is valuable lies buried in confusion, and covered with an almost impenetrable mass of worthless matter.

The 1832 cholera epidemic forever altered the medical landscape, creating problems for the medical profession that would be transformed by enterprising alternative medical sects into a professional crisis of epistemological proportions.

THE MAKING OF AN EPISTEMIC CONTEST

This chapter recounts how an epistemic contest developed out of the 1832 cholera epidemic. The epidemic disrupted the normal functioning of regular medicine. But in itself it did not cause allopathic medicine to reevaluate its intellectual foundations or question its professional future. Rather, alternative medical movements transformed the opportunities afforded by the epidemic into a crisis that forced allopaths to give an epistemological account of their knowledge. In the politics of knowledge that ensued, regulars' budding professional program was derailed, as state licensing laws, passed prior to the epidemic, were universally repealed. The intensification of competition in the newly unregulated medical market was joined with fierce debates over the nature of knowledge to produce an epistemic contest that would take nearly a century to resolve. Cholera may have entered the United States through a poor, unfortunate Irish immigrant, but alternative medical movements ensured that its humble origins belied its eventual impact.

When cholera arrived, the intellectual foundation of allopathic medicine was already in a fragile state. Rationalism—the intellectual foundation of allopathy—was coming under increasing scrutiny by some within allopathy who advocated for a more empirical approach to medical knowledge. Cholera exacerbated these internal tensions. Unable to provide a coherent picture of the disease, regular practitioners attempted to justify their professional authority, not on intellectual grounds, but on their standing as learned men. With deaths mounting, however, these epistemological debates, formerly latent and circumscribed within the profession, became public issues with life or death ramifications. Alternative medical practitioners, particularly Thomsonians and homeopaths, drew on the uncertainty introduced by cholera to force public medical debates onto the terrain of epistemology. What they offered were more democratic medical epistemologies. Epistemologies imply a social order (Shapin and Schaffer 1985). At the most basic level, they discriminate between those who are legitimate knowers and those who are not. From this basic distinction follows cultural (i.e., whose testimony is to be trusted) and organizational effects (i.e., who controls the institutional production of knowledge). Epistemic contests open possibilities for the reformulation of hierarchies in knowing. As this chapter shows, Thomsonians and homeopaths sought to undermine the traditional social order of knowing in medicine, by proffering more democratic visions for medical epistemology, which posited a role for the public in the production of medical knowledge.

To explain alternative medical sects' success in transforming cholera into an effective epistemological challenge, I embed these epistemological debates within the institutional contexts in which they unfolded, rather than conceiving them as unfolding in an abstract entity like the "public sphere." Epistemic contests do not occur in vacuums; they traverse the written page and enter into institutional and organizational contexts that shape their trajectories. Involved in sense-making (Weick 1979), organizations have internal cultures that shape the way information is understood, disseminated, and ultimately assessed (Vaughan 1996). In this way, organizations can be viewed as "epistemic settings" (Vaughan 1999) that delineate acceptable practices and procedures for the production and evaluation of knowledge. They are rhetorical spaces [that] "structure and limit the kinds of utterances that can be voiced within them with a reasonable expectation of uptake and choral support" (Code 1995, ix–x). I draw on the metaphor of an arena to make sense of the influence of organizations on epistemic contests. Arenas are defined by rules, more or less formalized, that shape strategic action—and influence outcomes—within them (Jasper 2006). Different capacities are needed to compete successfully in certain arenas, and therefore strategies must be designed to fit the context in which they are operating.[2] For intellectual disputes, actors must necessarily either forgo rhetorical arguments that are incongruous with that arena or lose.

The post-cholera medical debates turned on the issue of licensing and, as such, were situated in state legislatures. By the 1830s, regular physicians had begun to gain professional authority, successfully lobbying thirteen state legislatures to pass licensing laws (Numbers 1988). Yet, only a decade after the 1832 epidemic, these laws were universally repealed. Drawing on the insights of "new rhetoric," which links the success or failure of rhetorical arguments to the particular audiences and contexts that they address (Perelman and Olbrechts-Tyteca 1969), I examine the case of the New York State legislature in detail, to reveal the ways in which alternative medical movements' arguments—and the manner in which they were rhetorically presented—resonated with antebellum state legislatures. While regulars' hierarchal notions of medical knowledge clashed with the culture of the state legislature that was increasingly influenced by the ideals of Jacksonian democracy, alternative medical movements, seizing the democratic moment, promoted more egalitarian visions for medicine, which convinced the state legislatures to repeal the licensing laws and deregulate American medicine. In recounting this history, this chapter identifies the genesis of allopaths' problematic relationship with the state that would frustrate their professional goals and perpetuate the epistemic contest over medicine for nearly a century.

THE DECAY OF RATIONALISM AND THE CRUTCH OF AUTHORITATIVE TESTIMONY

In his 1833 presidential address to the Medical Society of the State of New York, Thomas Spencer summed up the regular profession's anxiety in the wake of cholera. Taking inventory of the ignorance surrounding the disease, Spencer (1833, 217) declared,

> Epidemics have in every age excited the dismay of mankind, and swept from the stage of human action a vast proportion of the inhabitants of the globe. The apprehension they produce is greatly enhanced by the rapidity of their movements, and the mysterious character in which these insidious enemies are enshrouded. It therefore becomes peculiarly important that the nature of every disease prevailing under this form, should be carefully investigated, and that the symptoms and mode of treatment found most successful, should be faithfully recorded.

This reasonable call for more research on cholera was sullied by the fact that, symptomatic of his peers, Spencer offered no clue as to *how* to answer these questions or what a "careful investigation" would actually entail. Instead, he goes on to dismiss cholera, not as something new, but as "a disease long known by the name of diarrhea serosa" (Spencer 1833, 218). In one quick stroke, Spencer explained away cholera's mysteriousness, claiming it was merely a variation of a familiar disease. A simple name change divested it "of mysticism" (Spencer 1833, 220).

Spencer's startling conclusion contradicted allopathic common sense. Upon what did he justify his unusual claim? On the one hand, Spencer suggests that it is built upon empirical observations, the "detail of the symptoms and the practical results to which my observations and investigations have conducted me" (Spencer 1833, 218). Yet such observations are never presented, nor is the nature of his investigations. This appears to be little more than a rote appeal to experience. On the other hand, he seems to situate cholera within a traditional rational system of disease (e.g., "Is it *rational* to believe, that diarrhea has its essential character changed, by becoming epidemic, and is thus rapidly disseminated by contagion?" [Spencer 1833, 288 emphasis added]). As to the components of this rational system, Spencer likewise remains silent. Without evidence to assess or a rational system through which to make sense of his claims, Spencer's declaration ultimately stands or falls on his own authority.

As founder of the Medical College of Geneva (New York), Spencer was considered one of "the most eminent physician of central New York," ("Death of Doctor Thomas Spencer" 1857, 6) and an important figure in allopathy. And while his conclusion may have been atypical, the fact that such a preeminent doctor succumbed to such muddled reasoning when encountering cholera underscores the epistemological problems facing regular medicine. When cholera arrived in the United States, the epistemological foundation of allopathy was languishing in ambiguity—torn between understated commitments to rationalism and jejune calls for empiricism. By 1832, the traditional foundation for allopathic knowledge—rationalism, or an approach to medicine by which particular cases were interpreted through universal, speculative systems of disease—was coming under criticism by regular reformers calling for knowledge rooted in bedside observation. Cholera intensified these calls, and the tensions between rationalism and empiricism, evident in Spencer's attempt to walk a fine line between the two, grew.

Given the uncertain foundation of medical knowledge, it is not surprising that during this period cholera was a truly heterogeneous thing, lacking a fixed identity and prone to multiple, often contradictory, interpretations. Regulars could not reach consensus on the most basic questions of the epidemic. Debate focused on three major issues, none of which would be resolved until decades later. These questions included:

- *Was cholera a new disease?* Despite its unusual symptoms and morbidity, many U.S. doctors doubted that cholera was something altogether new. The debate over cholera's identity focused on "whether this be a *new* species of *morbid action*, one *peculiar to itself*, or whether it be *similar* to the medical actions that obtain in other cases, in the same structures, and *differing* from them *only in degree*" (Hott 1832, 60). Complicating matters was the fact that the very *idea* of specific disease entities—that diseases had discrete causes and characteristic courses—was contested (Rosenberg 1987b, 72).
- *How was cholera transmitted?* Allopathic physicians "copiously argued in many a bulky library" (*BMSJ* 1835, 13) whether cholera was contagious or not. Sides were chosen depending on which contradictory evidence was stressed. Some noted cholera's movement along lines of travel, deeming it contagious (e.g., *BMSJ* 1832e, 254). Other reports, focusing on the isolation of cases and the lack of illness among medical professionals, declared it "to be wholly independent of contagion" (Comstock 1832, 353). Candidates for noncontagious causes proliferated: atmospheric influences (Clarke 1846), foul air (Comstock 1832), sudden changes in temperature (Williams 1844), and cosmic events like the alignment of the planets or an approaching comet (Allen 1832). Confusing the situation further, some doctors questioned whether the "distinction drawn between epidemic and contagious diseases was altogether fanciful. The fact is, that contagious diseases may become epidemic; and epidemic diseases, originally dependent upon atmospheric causes, may become contagious" (*BMSJ* 1831a, 14). As the distinction between contagious and noncontagious broke down, its contours became difficult to even outline.
- *How should cholera be treated?* Doctors threw their entire therapeutic kitchen sink at cholera. Among the suggested treatments were: traditional, heroic treatments like bloodletting, calomel, chloroform, opium, and emetics (*New York Journal of Medicine* 1849a); "sedative anti-spasmodics" (*BMSJ* 1831a, 13); saline (Sterling 1849); brandy and laudanum (*BMSJ* 1831a); hot milk and brandy (*New York Journal of Medicine* 1849b); wearing wool; and

> fleeing to the country (*New York Journal of Medicine* 1849a). Treatments would receive glowing reports in medical journals, only to be dismissed in the next issue.

Regulars could identify the relevant questions regarding cholera, but not *how* to answer them.

This confusion had deep origins in the undecided epistemology of allopathy. In pre-cholera times, the avoidance of epistemic questions was not much of an issue for regulars. This is not to suggest that allopathy lacked an epistemology in the early 1800s, just that it operated on an unspoken and unreflective plane. Typically, unarticulated epistemological commitments are not a problem, as these commitments are a tacit part of the taken-for-granted conventions surrounding knowledge and hierarchies in knowing. However, once the conventional ways are questioned and an epistemological challenge is mounted, the formerly tacit must be made explicit, and unquestioned assumptions must be justified. Cholera brought a sense of urgency to these internal epistemological debates and opportunities for the articulation of alternative epistemological visions by competing medical sects. People were dying; the public was losing confidence; and the long-held intellectual traditions of allopathy weren't helping. In the post-cholera world, regulars needed to articulate their vision for medical knowledge.

In the early decades of the nineteenth century, most allopaths retained a vague commitment to rational systems in making knowledge claims. Under rationalism, the diversity of diseases was reduced to a single (or at most a few) underlying cause, as physicians constructed elaborate speculative systems to make sense of disease (Warner 1997, 40). Ostensibly validated by experience, the "most striking feature" of these systems "was the rationalism that underlay their erection and operation" (Warner 1997, 41). These logical edifices of explanation made the practice of producing medical knowledge more akin to analytical philosophy than empirical science. Armed with these were "rigorously logical" (Shorter 1985, 30) abstract systems, allopaths attained knowledge of a particular case deductively by interpreting it *through* the lens of these systems. The most widespread rational system was developed by Benjamin Rush, who posited all local disease to be the result of vascular tension to be treated by depletive therapies (Duffy 1993). Other competing systems existed, most notably various humoral systems. Regardless of their specific differences, all rationalist systems shared the same orientation toward medical knowledge. Particular diseases were to be understood by inserting them into preexisting speculative systems. This is not to suggest that regulars who were committed to rationalism ignored empirical observation or experience altogether; rather their orientation toward observations was to interpret them through the lens of their particular rational system. Inconvenient facts, or those that could not be shoehorned into a given system, were treated as problematic anomalies, either ignored or set aside. Rationalism, thus, fostered a particular posture toward medical knowledge. Medicine was more an exercise in rational argumentation and logical deduction. Rationalistic accounts of cholera focused not on its particular manifestations, but rather on deducing how a given philosophical system could explain the disease. Did cholera represent an excess of bile? Was it an imbalance in the humors? A new manifestation of fever? Victory was won through philosophical speculation, not empirical searching or the presentation of data, as accounts were judged according to the logical argumentation displayed, with analogical reasoning comprising the bulwark of regular claims.

This method of adjudication through argumentation was not without its problems. Notably, arguments between incommensurable systems tended to end in stalemates. Because each system was built on its own unquestioned premises and assumptions, it was difficult to compare them. The assumptions that held *within* each specific system rarely held *across* different systems. What was logical in one was potentially inconceivable in another. This incommensurability was exacerbated by the fact that in journal articles and correspondence regular physicians often failed to explicitly lay out the systems they subscribed to, choosing instead to vaguely proclaim ideas "rational" or "irrational." Inevitably, this incommensurability made it difficult to compare and assess disputed claims about cholera.

Even prior to 1832, some regulars had grown ambivalent toward such rational systems. This ambivalence stemmed in part from frustration over the isolated islands of knowledge that the various systems created. Additionally, some worried that the rational systems, while elegant, could not capture the complexity of what they witnessed at the bedside. Patients with all their corresponding idiosyncrasies often did not fit into tidy rationalist systems. Finally, the ambivalence reflected a concern about the heroic therapies promoted by rationalist systems, like mercury and bloodletting, which "worked" in the sense that they induced visible and demonstrable physiological changes (Rosenberg 1987a, 74). Many allopaths became concerned that these extreme therapies were harmful and that rationalism encouraged their rote application to the patients' detriment and discontent (Warner 1997; Young 1967). These concerns, issued prior to the cholera epidemic, grew in its aftermath.

The initial turn away from rationalism was "by no means monolithic" (Warner 1997, 46), and calls for empiricism would not achieve coherence until later, when they were consolidated around the Paris School of medicine. Still prior to the epidemic, many were advocating a type of proto-empiricism whose parameters, although vague, were set in opposition to rationalism. Not yet a full-fledged, well-articulated epistemological system, proto-empiricism represented more of a general ethos or guiding principle that stressed the primacy of experience over logical argumentation in making knowledge claims. Its adherents were committed to inductive reasoning, reluctant to pigeonhole hard-won experience into a speculative system. This budding commitment to empiricism had roots in the "medical cosmology" of bedside medicine (Jewson 1974). In practice, bedside medicine stressed the interrelationship between the patient and the doctor built on familiarity gained over time, in which the local doctor had extensive knowledge of his patients. To treat disease, doctors discussed the symptoms of the patient and applied their wisdom to determine treatment (Lachmund 1998). Everything revolved around firsthand experience and observation. Proto-empiricism sought to make this practical technique of bedside observation the foundation for medical knowledge.

The essential intellectual component of this proto-empiricism was the doctrine of specificity (Warner 1997).[3] This principle claimed that a disease could only be understood by taking into account the idiosyncrasies of the (1) patient and (2) the region in which the disease occurred. Disease was seen as polycentric and polymorphous, varying across individuals and contexts (e.g., disease in New York was qualitatively different from disease in Georgia). Given such variation, doctors had to tailor diagnoses and treatment to the specific case. This epistemological stance contrasted greatly with rationalism, as it caused a reluctance to universalize and draw analogies between disparate geographic regions or even between two different patients. Because knowledge was understood as specific and localized, empirical allopaths were not particularly concerned with accumulating particular facts so as to achieve a universal explanation of disease. Medical knowledge was oriented toward the exigencies of treating a particular patient, not toward achieving universal, abstract knowledge. These doctors refused to engage in the type of philosophical speculation rampant under rationalism. In fact, what proto-empiricism offered was less a positive program for the future of medical knowledge, and more a critique of past rationalism, a negative program of "tearing down" the troublesome speculative systems (Warner 1997, 59). As such, while the early commitment to empiricism was still without a cohesive vision, it nonetheless offered a firm position from which to critique rationalism.

Torn between rationalism and proto-empiricism, allopaths lacked a clearly articulated epistemological foundation, which produced a fragmented knowledge base and internal discord over the most basic assumptions regarding medical knowledge. Under periods of business as usual, such fragmentation was not too damaging. But cholera changed this, as it increased the public stakes of these debates. It forced these debates, formerly circumscribed within allopathy, into the light of day. Such fragmentation led not only to inconsistent accounts of cholera but also to muddled interventions and ultimately death.

In addressing their disunity, regulars retreated to claims of authority based on their status as learned men in the community. Regular professional identity during the early nineteenth century was not built on a shared, coherent body of knowledge, but on a shared sense of status, derived more from common therapeutic practices than a coherent system of thought (Warner 1997). All regulars, regardless of their epistemological commitments, subscribed to a similar conception of the hierarchy in medical knowledge that situated themselves at the apex of authority on medical matters and denied the legitimacy of knowledge emanating from those outside of the regular community. And both rationalist and empiricist allopaths rooted their knowledge claims in their positions of authority in the community; their status as learned men required that their views be privileged. For rationalists, this authority rested on their access to a tradition of philosophical medical thought. For empiricists, it was grounded in a notion of doctors as superior observers. Regardless of its ultimate foundation, this was a type of "generalized wise man" model (Parsons 1991, 295) of authority, based more on *who* regulars were than on *what* or *how* they knew. Thus, while never very extensive during this period (Shorter 1985), the deference allopaths received from the public was based upon local reputations as learned men and moral citizens. Respect was demanded in a manner akin to the way that local communities confer trust and authority to the ideas of clergy and other learned men. Upon this shared identity as members of the learned elite, regulars attempted to unite despite their disparate epistemological orientations.

Grounding their authority on their reputations, regular physicians felt little compulsion to justify or explain themselves to the lay public. Instead, they resorted to *authoritative testimony* in communicating knowledge. Regulars made claims through dogmatic assertions of personal and traditional authority. They refused to offer accounts for their assertions, to present data that supported their claims, or to entertain competing ideas from alternative sects. Rather than lay out their reasoning, regular physicians stated their conclusions, offering only the barest of justifications. "Facts" were not presented but proclaimed. Their status spoke for their competence, their privilege derived from the source of knowledge—who was proclaiming it—not on the content. Regulars' use of Latin in defining medical terms and composing treatises reflected the underlying assumptions of hierarchy in knowing to which allopathy subscribed. Because of its inaccessibility and opacity, Latin demarcated legitimate knowers from non-knowers. Knowing the language of medicine signified membership in an elite community. If one could not participate in such Latinate discourse, one was not meant to meddle in medicine. Consequently, authoritative testimony as a rhetorical strategy served to mystify knowledge in such a way that masked the epistemological uncertainty permeating the profession. It allowed regulars to *tell* rather than show. In essence, they covered up their lack of epistemological coherence through strategies of avoidance, by simply refusing to discuss them, and/or by deflection, shifting these concerns onto issues better handled by their rationalist systems.

Still, the denial of epistemological issues and refusal to engage with these basic issues had corrosive effects and presented problems for allopaths in their attempts to make sense of cholera. Authoritative testimony prohibited regulars from discriminating between legitimate accounts and spurious ones *within* their own sect, a persistent epistemological problem that plagued allopathy throughout the nineteenth century. There was no way to reach any consensus on cholera with knowledge that was based primarily on the reputation of the knower. For rationalists, authoritative testimony allowed them to gloss over contradictions in their competing rational systems, thereby prohibiting the type of hard-won consensus the public clamored for. Since they simply stated claims, rather than showing their reasoning, there was little to actually assess. The criteria and logic of any testimony remained invisible. So while masked, the conflicts and contradictions between rationalist systems went unresolved; regulars committed to rationalism presented a unified front, but in actuality confusion reigned. Unsurprisingly, this confusion found its way into the proclamations on cholera.

For empiricists who had discarded the traditional systems and replaced them with an empiricism that put a premium on experience, authoritative testimony did little to solve the problem of communication beyond the local context. Traditionally, testimonies would be assessed according to the reputation of the testifier, as doctors drew on their familiarity with their peers' reputations to make determinations of trustworthiness. Proto-empiricists lacked the standard or techniques to assess the validity of experiences as testified to by others in situations where personal familiarity with the knower was lacking. This was especially problematic for a disease like cholera that traversed local contexts. Cholera was a disease born of the structural changes in the first half of the nineteenth century. Revolutions in transportation and communication overthrew the "tyranny of distance" (Howe 2007, 225), as did rapid urbanization. Natural barriers to international interaction like the Atlantic Ocean were now easily traversed by both people and diseases. Inland, the emergence of a national market system, organized around canals and railroads, facilitated the diffusion of cholera throughout the United States. This movement was painstakingly recorded by the press, which grew threefold in the decades between 1820 and 1850 (Reynolds 2008), following the same transportation revolution as cholera itself (Mindich 2000, 96). As people's horizons grew with increased interactions across local contexts (Haskell 1985) and rapid urbanization (Howe 2007), they became more familiar with what was going on beyond their local communities and began to think in terms of larger, more impersonal collectives. Thus, these very changes, which enabled cholera to spread, also undermined the traditional authority of doctors. Local personalism and trust based primarily on familiarity no longer held in this new environment (Halttunen 1986). The translocal character of cholera not only compromised allopathic locally rooted authority; it also threw the inadequacies of its epistemological foundation into stark relief. Because allopathy lacked a well-articulated *general* program for medical knowledge, it struggled to make claims and assessments that traversed the local. Without any universal standard or measure of good knowledge, and with little information beyond individual declaration offered as proof, there was no way to tell a useful testimony from a useless one, no method to communicate effectively across localities in order to develop a clear, comprehensive picture of the disease.

The end result of this epistemological discord was a lack of coherence in accounting for cholera. In the face of such confusion, regulars dug in their heels, leaning more and more on authoritative testimony. Rather than reaching conclusions on cholera, allopathic journals produced little more than incessant testimony and irresolvable debates. For example, take the endless, circular discussion on treatments for cholera. In an 1832 article in the *Boston Medical and Surgical Journal*, Dr. Robert Lewins (1832, 273) argued that a cure "may be accomplished by injecting a weak saline solution into the veins of the patient. . . . The most wondrous and satisfactory effect is the immediate consequence of the injection."[4] In the very next edition of the same journal, Dr. P. Bossey (1832, 245), however, offered the opposite claim; saline injections were found "totally inert and injurious." Yet another doctor was told by an English doctor "that common table salt—a spoonful in a tumbler of water—not only speedily relieved him [the patient] from violent pains, but ultimately restored him to perfect health" (*BMSJ* 1831b, 170). Every author supported his claims with personal testimony which was inevitably followed by counterclaims. With no way to assess these competing claims, with no evidence beyond individual testimony, regular medical knowledge devolved into an endless proliferation of divergent claims, and the public was left confused as to what it all meant.

DEMOCRATIZING MEDICAL KNOWLEDGE

The use of authoritative testimony did not resolve internal epistemological debates within allopathy, but it did mask them via the omission of evidence of these debates in allopathic journals and meetings. However, the attempt to justify knowledge on authority, problematic *within* the profession, was even more dubious *outside* of the profession, as alternative medical movements began to demand an epistemological account from allopaths to justify their professional privileges. While regulars sought to avoid epistemological conflicts, alternative medical sects afforded them no such opportunity, leveraging the epidemic to force an epistemic contest over medicine.

In part, the challenge to *medical* authority reflected the tenor of the times, in which challenges to authority were becoming rife throughout American culture. During the early nineteenth century, "the politics of assent" of the previous era gave way to the rough and tumble world of political parties (Schudson 1998). Democratization touched all sectors of American society—in religion through the Second Great Awakening, in reform movements like temperance and abolitionism, and in government with the rise of mass political parties and the extension of voting rights. And the growth of a vibrant press literally made knowledge more available to the public (Schudson 1981). Anti-intellectualism accompanied Jacksonian democracy, which "completed the disestablishment of a patrician leadership that had been losing its grip for some time" (Hofstadter 1963, 51). Cholera arrived precisely at the time that these democratic trends were crystallizing into conditions ripe for an epistemic confrontation in medicine. The crutch of regular testimony solved some problems, but it faced an increasingly hostile environment. Authoritative testimony declared rather than showed, leading to problems for regulars when entering public institutions that asked them to engage in debate by *showing* their reasoning.

Alternative medical movements seized these trends and married them to critiques of regulars' handling of cholera to mount a forceful attack on allopathy's professional program. Cholera provided an opportunity for alternative medical movements that, in different ways, sought to democratize knowledge. The lesson of cholera was clear enough to the public; regulars' heroic therapies inflated the mortality of the disease (Rosenberg 1987b, 72). Many turned to the milder alternative medical sects, and their ranks swelled. Regulars lamented these conversions, complaining that the public "through sheer ignorance of the steps leading indirectly to the temple of science . . . swallows with avidity the monsters of quackery *practice*" (*BMSJ* 1831a, 13). Regular physicians' concern for their professional status grew, as did their sense of panic. By 1835, Edward Deloney (1835, 111) complained, "At no period of the world, even in the dark ages of superstition, has the profession of medicine been thronged with imposters of the most daring effrontery than at the present time."

The failure of regulars to stem cholera, or even provide a coherent account of it, became a crucial symbolic resource for alternative medical sects. Armed with alternative framings of cholera, these sects took aim at regular therapeutics, their use of authoritative testimony, and their privileged professional position. But most important, the cholera epidemic offered an occasion for alternative medical movements to force medical debates onto the level of epistemology. By offering competing medical epistemologies resonant with increasing democratic sentiments, these alternative medical sects made allopathy's elitist orientation seem retrograde.

Thomsonism: Every Person, His or Her Own Doctor

The most radical attempt to democratize medical knowledge arose out of rural New Hampshire. Samuel Thomson, a combative, somewhat paranoid, and yet charismatic farmer, suspected regulars of withholding information from the people so as to garner power and wealth. In response, he developed his own herbal system of medicine—Thomsonism—that was simple enough for anyone to use. His vision was nothing less than a system of medicine without doctors.

When he first published his *New Guide to Health* in 1822, Thomson saw his system as the defender of true medical knowledge—knowledge that was available to all. Under the Thomsonian system every person was to become his or her own physician (Kett 1968).[5] The antiauthoritarian and even conspiratorial character of Thomson's critique fell on fertile soil during the Jacksonian era, and his system spread throughout the country via a network of itinerant "healers." It caught on, especially in poorer, rural areas (Coulter 1973, 92), growing through the careful exploitation of popular sentiments, egalitarianism, nationalism, and romanticism (Whorton 1982, 24). Although statistics from the era are notoriously imprecise, historians believe that during its height in the 1830s, Thomsonism claimed over a million followers (Berman and Flannery 2001; Haller 2000). In 1833, a Thomsonian magazine published a list showing authorized agents in twenty-two states and territories (Rothstein 1992, 45). By 1840, Thomson had sold one hundred thousand family rights, and approximately half of the citizens in Mississippi and Ohio were curing themselves using Thomsonian methods (Numbers 1988).

The Thomsonian system by no means offered a revolution in medical ideas. In some respects, it was even more traditional than the medical system practiced by regulars, committed as it was to the type of rationalism that allopathy was in the process of discarding (Kett 1968). Drawing on the humoral system of disease, Thomsonians held that all illness arose from an imbalance of heat in the body. Tracing all disease to this single cause, Thomson believed that effective therapy required increasing the body's heat through natural remedies like botanic medicine and sweat baths. Rather than combating nature through depletive therapies like regulars, Thomsonians viewed nature as inherently ameliorative and used it as their guide. In turn, they employed milder treatments with fewer side effects than allopathy's use of calomel and bloodletting—two practices Thomsonians viewed as "unnatural and injurious" (Thomson 1825, 206). The milder treatments served them well during the cholera epidemic as they were less harmful than allopaths' extreme heroic treatments.

Although conventional in its medical ideas, Thomsonism aspired to revolutionize the relationship between the physician and the patient by exploding all hierarchies in medical knowledge. Foremost, the Thomsonian challenge was an epistemological one that denied authorities in medical knowledge. Appealing to egalitarianism, it stressed common sense and the common man as a knower (Haller 2000). The *Boston Thomsonian and Lady's Companion* (*BTLC*) (1840, 338) championed the medical potential of folk knowledge: "The exercise of our reasoning faculties at once puts us in the right road to discover truth." Thomson intentionally developed his system to be used by anyone, regardless of education. The system was written and distributed in an accessible handbook that anyone could purchase for a small price and subsequently use to treat his or her family. And Thomson promoted his system through stories, popular lectures, and poems that served as mnemonic devices (Haller 2000). Every individual could digest medical information presented in these forms. Everyone could practice medicine. Doctors were not needed.

Thomson put faith in the agency of the patient to such a degree that the distinction between doctor and patient would be eliminated. Thomson's own narrative embodied the ideals of his system, and he appended it to the numerous editions to his *New Guide to Health*. Uneducated himself, Thomson discovered the herb *Lobelia inflata* when he was four, his consumption of which quickly induced vomiting. While he did not yet recognize the significance of his discovery, this herbal emetic would become central in his therapeutic arsenal. In 1788, at the age of nineteen, he badly injured his ankle and after enduring the painful failures of regular therapeutics, treated himself successfully using folk knowledge culled from experience. In doing so, Thomson was simultaneously the patient and the doctor—an arrangement he later championed in his system. He began treating his family and neighbors, and as his reputation grew, he abandoned farming to become a full-time practitioner. The remainder of the narrative recounts Thomson's healing successes in the face of hostility and challenges from regulars. The lesson of his narrative is explicit: the talent for healing is not restricted to the profession; it is in everyone.

According to Thomson and his followers, regular physicians, acting out of avarice and self-interest, would stop at nothing to prevent the public from learning this revolutionary lesson. In making medical knowledge available to all, Thomsonians juxtaposed themselves to regulars, who tried to maintain a monopoly over medical knowledge by obscuring its simplicity (Haller 2000). Education was a tool that regulars used to monopolize and mystify medical knowledge. Sarcastically challenging the intellectual pretensions of allopathy, the *Thomsonian Messenger* (1843, 74) dismissed regular educational elitism:

> We admit that we are not so wondrous wise as some of our would-be medical Solomons, nor can we so readily mouth the vocabulary of learned technicalities, as some of the M.Ds. But we consider that to understand LIVING ANATOMY, and how to KEEP ALIVE, is of infinitely more importance to both the patient and practitioner, than weeks, months, or even years spent in comparatively useless studies, or in shaking the dry bones of the human skeleton.

Thomson himself (1825, 199–200) argued, "The practice of the regular physicians, that is those who get a diploma, at the present time, is not to use those means which would be most likely to cure disease; but to try experiments upon what they have read in books, and to see how much a patient can bear without producing death." Furthermore, the regulars sought to confuse patients and obfuscate their inability to cure by using Latin, which concealed knowledge "in a dead language" (Thomson 1825, 193). Regulars "have learned just enough to know how to deceive the people, and keep them ignorant, by covering their doings under a language unknown to their patients" (Thomson 1825, 41). The secrecy of the regulars was opposed to the transparency of the Thomsonian system, which "so far from concealing discoveries or seeking to make a mystery of them" labored "to make them known for the benefit of the whole human race" (Thomson 1839, 50). Such a commitment to openness was on display during the 1832 epidemic, as Thomsonians made their anticholera recipe widely available to the public (Haller 2000).

Contra allopaths, Thomsonians believed that relevant knowledge was not found in books, but in the common sense of the people. Folk wisdom was prized over education. Thomson (1825, 34–35) wanted to unlock the inherent, good sense of the people, hoping

> that it (the Thomsonian system) will eventually be the cause of throwing off the veil of ignorance from the eyes of the good people of this country, and do away with the blind confidence they are so much in the habit of

placing in those who call themselves physicians, who fare sumptuously every day; living in splendour and magnificence, supported by the impositions they practice upon the deluded and credulous people; for they have much more regard for their own interest than they do for the health and happiness of those who are so unfortunate as to have anything to do with them.

Underlying this egalitarian epistemology of common sense was a view of truth as transparent and acceptable, as nature provided the requisite clues for those careful enough to observe. The notion that truth was mysterious was anathema to Thomsonians: "Truth never seeks to be sheltered in mystery, she delights in simplicity, because it adorns her by laying all her beauties open to general inspection" (*BTLC* 1841, 114). Nature made itself readily comprehensible to all. The public therefore had the capacity to grasp medical knowledge. The flattened hierarchy of knowers and egalitarian Thomsonian epistemology was defined in opposition to the authoritative, paternalistic approach of regulars, who mystified medical knowledge for selfish ends to the point where "the acquirement of medical knowledge having been considered a matter beyond the ordinary pursuits of life, the people have fallen into a complete state of darkness and superstition on this important subject" (Colby 1839, 2). If the public shed the belief in the authority of the regulars, medical insight was there for the taking. By juxtaposing their epistemology of folk wisdom and common sense with allopathy, Thomsonians demanded epistemological account from allopathy that did not fall back on obfuscation.

Cholera offered a powerful symbol of regular failure that Thomsonians exploited to undermine licensing laws. As part of their rhetoric, they consistently contrasted the self-proclaimed success of their therapeutics with the dismal record of the regulars. Thomsonian periodicals were rife with condemnations of regular therapeutics like calomel and contrasting stories of the successes of Thomsonism. Bloodletting was viewed as not only ineffective, but deadly in treating cholera (Thomson 1825), as regulars "destroy more frequently than they can save" (Whitney 1833, 319). As for their own success, Thomsonians claimed, in a fit of hyperbole brash even for the period, to have saved 4,978 of the 5,000 cholera patients they treated in New York City (Haller 2005, 98). The critique of the allopathic handling of cholera was a common refrain among all alternative medical movements. The innovation Thomson brought was to link this failure to the monopoly regulars held over medicine, reframing the tragedy of cholera as an exemplar of what happens when a profession with monopolistic power cannot be held to account. The cruelty of heroic medicine in treating cholera was a direct outcome of licensing laws that encouraged callous experimentation. These laws were framed as part of a concerted campaign by regular physicians to stifle freedom in a way that violated the spirit of American democracy: "Exclusive privileges are incompatible with those for civil rights for which the martyrs of the revolution fought and bled and died!" (*Thomsonian Botanic Watchman* 1834a, 131). Furthermore, the monopolies attained through licensing laws created an apathetic public, stunting their intellectual capacities and, in turn, stifling medical progress. An editorial in the *Boston Thomsonian Manual* (1841, 98) noted the pernicious effects of monopolies on the intelligence of the public: "Every individual who permits himself to be ignorant of medical knowledge, exposes his life to a banditti, who are 'regularly' drilled and legalized to go out into the world and experiment on human life. It is a duty we owe our God, ourselves and families, to exercise our reasoning faculties, in order that we may not only benefit ourselves but those of our species." Because the talent for healing could develop in anyone regardless of educational attainment, licensing laws were not just inefficient; they were immoral.

In promoting their vision of a democratized medical epistemology, Thomsonians often sounded more like a political and economic movement for egalitarianism than medical reformers. Indeed, the great innovation of Thomsonism was to take the rhetoric from democratizing religious and political movements and apply it to medicine. John Thomson (1841, 172), Samuel Thomson's son and ardent disciple, made this connection explicitly: "People who are competent to judge who shall be their Legislators, are also equally qualified to select their doctors." Prohibiting people from exercising their ingrained intelligence was an unspeakable evil, a "tyranny for a free country!" (*Thomsonian Botanic Watchman* 1834b, 86). It was also, according to Thomsonians, bad medicine.

Homeopathy and the Democratic Rhetoric of Statistics

Homeopaths adopted a subtler approach to democratization. Rather than make every person his or her own physician, homeopaths retained their position of expertise but invited the public to assess their knowledge claims. As such, they championed an active role for the public that contrasted with allopathic elitism without succumbing to what they perceived to be the radical excesses of Thomsonism. The manner in which the homeopathic approach to democratizing medical knowledge diverged from Thomsonism reflected the different clientele of each sect. Unlike the anti-intellectual Thomsonians, homeopaths were well educated, growing their ranks by luring converts away from allopathy (Kaufman 1988). Homeopathy was a decidedly cosmopolitan medical system, strongest in the urban centers of New England, New York, Pennsylvania, and the Midwest (Coulter 1973, 108). Homeopaths' high level of education, their scientific bona fides, and their patronage by urban elites facilitated their rise. By wielding a conscious and explicit epistemological vision that laid claim to the mantle of medical science, homeopathy called allopathy to account in a way that the uneducated, rural Thomsonians could not. As such, they would become the most persistent and successful challengers to allopathy (Coulter 1973; Kaufman 1988).

Homeopathy was imported to the United States in the 1820s, winning significant American converts after the perceived success of its milder treatments during the 1832 cholera epidemic. Samuel Hahnemann, the founder of homeopathy, was a German physician who became disillusioned with regular medical practice, arguing that it did more harm than good. In opposition to the heroic practices of his peers,[6] Hahnemann developed a milder, less intrusive form of medicine.

Hahnemann viewed disease as an imbalance of the vital force. As Hahnemann conceived it, the vital force encompassed physical, mental, and spiritual properties, and disease represented disequilibrium in any of these properties. This would seem to place homeopaths in a more traditionally rationalist camp. But this was not the case. While the vital force served as the philosophical underpinning of the homeopathic system, it was considered essentially unknowable. Given the vital force's inscrutability, homeopathy adopted a posture toward medical knowledge quite distinct from rationalist system building. Disparaging the search for disease etiologies as misguided, homeopaths focused instead on the rigorous empirical investigation of physical manifestations (i.e., symptoms) of how the vital force was responding to disease. Homeopath C. L. Spencer (1857, 9) outlined the homeopathic method: "The real fundamental principle of vital life, the great motive power of our existence, never has been seen, nor never will be seen; none but the Supreme Ruler of the universe can fathom this mystery. But we have symptoms in all of their various phases left us for our guide; each disease develops symptoms peculiar to itself." The vital force left crucial clues that when attended to could inform homeopaths of how to help it heal. Thus, in a counterintuitive way, homeopathy's mystical understanding of the vital force did not devolve into speculation or mysticism, but rather promoted an ethos of rigorous scientific empiricism.

Such empiricism was reflected in the major contribution that Hahnemann claimed for his system—*similia similibus curantur*, or law of similars. According to this law, disease symptoms, rather than reflecting the pathological effects of disease, revealed the body's efforts to bring the vital force back into equilibrium. Homeopaths claimed the law of similars as scientific; it was "discovered" by Samuel Hahnemann through a careful consideration of facts, for "nature's laws are ascertained in just one way—by observation" (Hand 1874, 18). The goal of homeopathic therapeutics was to assist the vital force toward equilibrium, by providing treatments that induced similar symptoms. While the story of Hahnemann's discovery may be apocryphal, it speaks to the homeopathic commitment to observation in illuminating universal laws. Treatments were determined by matching the symptoms induced by these treatments to those that the patient was experiencing. Building a therapeutic system demanded the accumulation of fine-grained observations of symptoms. Consequently, the recognition of the limits of human reason led to a commitment to empiricism and induction as a means of uncovering facts (Coulter 1973). Francis Hodgen Orme (1868, 35), a southern homeopath with expertise in yellow fever, claimed, "The logic of the homeopathist, then, is the logic of facts." It was in their embrace of facts that homeopaths staked their claim to science. "The present age demands, and rightly, that no element of medical science, of which we possess the means of gaining positive knowledge, shall any longer be left to conjecture. It demands facts," argued homeopath P. P. Wells (1864, 91). Accumulated facts would lead homeopaths to laws regarding the underlying uniformities in nature. This commitment to induction and observation was reflected in their experimental orientation, whereby insights were gained through tests of nature. They determined treatment regimens by conducting experiments they called "provings" in which a single drug was given to a healthy individual, often the doctor himself, who was carefully monitored for symptoms by taking elaborate case histories.[7] The goal was to accumulate a large body of knowledge on the symptoms induced by a given medication in order to later match the symptoms of the ill patient to the symptoms induced by the drugs in the healthy person.

In no area was homeopaths' commitment to gathering facts more evident than in their use of statistics. Unlike regulars, homeopaths accumulated a mass of detailed quantitative data to discover the specific drug remedy for the totality of symptoms experienced by a patient. Managing all this data required a sophisticated understanding of numbers. Homeopathic editors regularly solicited statistical reports, while homeopathic societies collected statistical data on diseases in their region (Cassedy 1984, 126).

Statistics have rhetorical dimensions (Carruthers and Espeland 1991); they are "figures of speech in numerical dress" (McCloskey 1985, 56) used by actors to gain an argumentative advantage through the perception of their disinterestedness and sober objectivity. Both allopathy and homeopathy lacked a coherent explanation for cholera, but the homeopathic use of statistics conveyed a confidence and measure of control that stood in marked contrast to the fuddled accounts of allopathy. Homeopaths used the *rhetoric* of statistics with aplomb as a tool of persuasion to promote a particular interpretive framing of cholera. Statistics reduce complexity (Starr 1987, 40), and homeopaths' statistical rhetoric transformed the messy ignorance toward cholera into a single number—the mortality ratio of the different systems used to treat it. Frederick Hiller (1867, 11) offered a typical framing of cholera by homeopaths. After presenting tables of comparative statistical data, Hiller argued, "The numerous, authentic, statistical reports from 1831 to 1863, from various countries, give the following comparative difference of mortality in this fearful disease: Allopathic mortality, 57 per cent; homeopathic mortality, 9 per cent." Cholera was reduced to comparative data points—statistical rates of mortality of the two systems—that were presented as self-evident. The data behind such statistical comparisons was suspect for a number of methodological reasons; this was not a case of homeopaths having superior information. But the rhetorical functions of statistics did much to bolster homeopaths' standing, especially in state legislatures. Through "uncertainty absorption" (Carruthers and Espeland 1991, 57), this statistical reduction suggested a degree of control over the phenomenon lacking in regulars' meandering discourse on cholera's etiology, while also providing homeopaths with information that traveled well across contexts. Homeopathy took the confusing information regarding cholera and repackaged it into a single comparative statistic that they wielded against regulars.

Underlying the rhetoric of homeopathic statistics was a social order of knowing inconceivable to allopaths. While the accuracy of the homeopathic statistics is difficult to measure—and indications are that they were generously inaccurate—they reflected an open and public epistemology based on demonstration. Present-day criticisms about lying with statistics aside, during this period, statistical analysis held the promise of solving a crisis of trust brought about by modern society (Porter 1988,1995). Statistics, and numbers generally, were understood in the nineteenth-century context as preevaluative and noninterpretive (Poovey 1998). As the boundaries of the local were transgressed and anonymity rose, statistics were seen as a way to make knowledge transparent, solving the problems of trust created by an increasingly impersonal society. While the United States lagged far behind Europe in the development of sophisticated statistical techniques, statistical thinking and numeracy in general assumed a prominent role in everyday American life (Cohen 1982). In his observations of U.S. society, de Tocqueville (2002 [1835], 285) found Americans' minds to be "accustomed to definite calculations." Thus, statistical data came to assume great importance during the Jacksonian period, as it came to be viewed as essential for good government and achieving consensus in a country in which numeracy was on the rise (Cohen 1982). Homeopathic statistics accommodated these structural and political changes, resonated with the shifts in governance and, in turn, stood in stark opposition to the traditional modes of authority of regulars.

Statistical rhetoric induced trust through participation. Data was presented to the public in a manner much different from allopathy's appeals to authoritative testimony. Homeopaths embraced the identity of Baconian scientists who gathered facts and laid them before the public to judge. They presented their findings in tables, enabling the public to assess their claims, asking them to serve as discerning judges in a manner that respected public opinion. For example, Edwin Miller Kellogg (1872, 2), after giving his statistical report of mortality rates in New York City for 1870 and 1871, concluded, "These statistics and the logical inferences therefrom we leave to the consideration of the thoughtful public." Rather than *dictating* knowledge like allopaths, homeopaths *demonstrated* claims through the presentation of facts. In reducing complexity to numbers and then presenting numerical data, homeopaths invited the public to assess medical knowledge. This respect for lay knowledge was built into the practice of homeopaths, as patients were expected to assume an active role in their treatment. Consequently, while homeopaths identified as Baconian scientists, their expertise was built on a cooperative, rather than didactic, interaction with the lay public, who were accorded the right to judge homeopathic claims. They embraced an identity founded on democratic ideals, for as homeopath Arthur Lippe (1865, 26) decreed, "The Homeopathician represents the true democratic principle in the healing art, he courts inquiry and lays facts before the people by which they may judge of the validity of his claims to superiority."

LICENSING REPEAL AND PROFESSIONAL COLLAPSE

The visions of medicine proposed by Thomsonians and homeopaths differed in the degree to which they sought to flatten hierarchies in medical knowledge (see table 1.1). Nevertheless, both offered a more democratic epistemology than allopathy. And in articulating competing epistemological visions and demanding an account of regular medical knowledge, Thomsonians and homeopaths shifted medical debates over particular issues, like cholera, onto the level of epistemology. By provoking comparisons between sects, these alternative medical sects suggested the possibility of choice in medicine. Seizing the opening afforded by the epidemic, they transformed cholera into an epistemic contest.

As argued above, epistemic contests play themselves out in institutional arenas, whose natures affect their trajectory. After the first cholera epidemic, no arena was more significant than state legislatures, where medical sects vied for authority through licensing laws. Prior to 1832, regulars looked optimistically to their professional future, having gained a measure of professional recognition with the passage of licensing laws in thirteen states (Numbers 1988). Laws varied across states but generally they either prohibited the practice of medicine by nonlicensed physicians (punishable

Table 1.1. Contrasting styles of reasoning among nineteenth-century medical sects

Medical sect	Nature of truth	Acquisition of knowledge	Dissemination of knowledge	Social order of knowers	Framing of cholera
Allopathy	Abstract and theoretical OR local and specific	Logical reasoning OR proto-empiricism	Authoritative testimony	Hierarchal, regular physicians only	Confused; lack of consensus
Thomsonism	Transparent in nature	Common sense and folk wisdom	Network of botanic societies sharing experiences	Egalitarian, every person his or her own doctor	Cholera as a failure of medical monopoly
Homeopathy	Fundamentally unknowable but glean clues through observation	Detailed observation of manifest system and particularities of the patient	Statistical rhetoric and demonstration	Hierarchal, but public as judges	Cholera as a number

by fine or imprisonment) and/or prohibited nonlicensed practitioners from suing for fees, an economic prohibition that could cripple a practice. Licenses were granted by medical schools, societies, or board of examiners, all bodies under allopathic control. While these laws were mainly symbolic, as it was difficult to enforce them (De Ville 1990; Kett 1968; Shryock 1967; Haller 2000), they had important symbolic value, for despite their limitations, they endorsed allopathy's claim to a privileged professional position. Allopathic optimism dissipated after the 1832 cholera epidemic, as the Thomsonians and homeopaths used regulars' cholera failure to advocate for the repeal of such laws. Rapidly, the regulations were repealed. Beginning with Alabama and Ohio in 1833 through the final elimination of all regular legal privileges in New York in 1844, states deregulated medical practice, the exceptions being New Jersey and the District of Columbia.[8] The repeals both humbled the profession and ushered in a period of intense competition between medical sects (Numbers 1988). By 1848, the arrival of the second cholera epidemic in

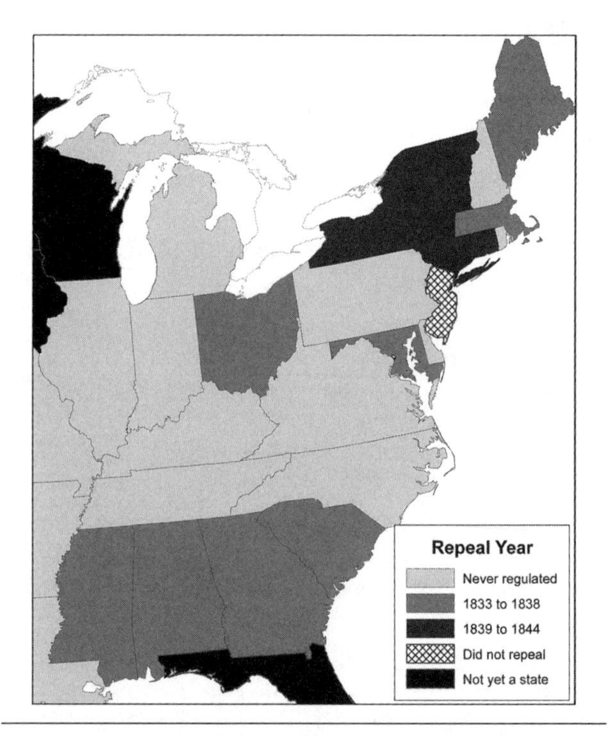

Map of repeal of state licensing laws. Mark Treskon.

the United States, government regulation of medicine had reached its "nadir" (Kett 1968, 27).

How did alternative medical movements achieve such success in their campaigns against licensing? By examining the legislative debates over licensing laws in the New York State legislature—the state with the longest and most protracted struggle—we can gain a general sense of how these competing rhetorics and epistemic visions played out in the state legislatures. In the New York State legislature, alternative medical movements' vision of knowledge found an audience generally receptive to their democratic ethos and hostile to regulars' appeals to authority. Thomsonism's elevation of common sense and homeopathy's recognition of the public as legitimate judges shared affinities with the culture and organizational practices of the legislature, which, despite its failings and contradictions, had been influenced by Jacksonian democratic ideals. In the end, Thomsonians and homeopaths offered something more substantial to the legislatures than better efficacy in addressing cholera. They offered a new vision of medicine that democratized knowledge.

Deregulation in New York

During the Jacksonian era, American political institutions underwent a democratic transition (Pessen 1969), as political reforms expanded suffrage and sought to make government more responsive to the people (Kass 1965). In the new era of mass political parties (Schudson 1998), politicians had to placate a growing electorate, if not in substance, at least in rhetoric. Certainly within the political realm, the rhetoric of democracy and the common man became common currency among all political parties and actors. Historians have long debated the extent to which the egalitarianism as espoused by politicians during this period was genuine (see, for example, Feller 1990; Kohl 1989). Even if the appeal to the common man was cynical, it provided an important cultural trope used by both political parties, Whigs and Democrats (Kohl 1989; Wilson 1974). The use of such rhetoric—the instrumental and, yes often, cynical appeals to democracy by politicians—reflected a substantive shift in political calculus toward a participatory role for the public, and politicians adapted accordingly, clamoring to present themselves as attentive to public input.

By 1832, democratic reform had already reached the New York State legislature. In the 1820s, New York expanded the franchise by repealing property qualifications for voting. Throughout the 1830s and 1840s, Jacksonian Democrats, organized under Martin Van Buren's Albany Regency, controlled the legislature, but the Whigs maintained a significant minority status. Regardless of who was in control, however, the similarities between the parties outweighed their differences (Benson 1961; Kass 1965). An emerging consensus among historians is that Whigs and Democrats shared a commitment to key unquestioned values—the permanence and immutability of the Constitution, the defense of liberty and republicanism as the prime goal of political activity, and the liberal economic precepts of private property and free enterprise (Feller 1990). More important, both political parties sought to gain mass appeal, differing only in their conceptions of and means to individual freedom, not in the significance of freedom as an ideal (Wilson 1974). Both learned to act and talk within a democratic universe, where "everyone loved the people, bowed gladly to their sovereignty, celebrated their virtue and their judgment" (Meyers 1957, 257).

While the actions of politicians rarely lived up to their professed democratic ideals, they did shape how the legislature viewed knowledge. The appeals to the common man by both New York Whigs and Democrats implied a view of knowledge production that advocated common sense and a mistrust of authoritative accounts. De Tocqueville (2000, 512) was struck by Americans' self-reliance on intellectual matters, their suspicion of authorities, and their continued appeals to common sense, noting "each American appeals to the individual exercise of his own understanding alone." The Jacksonian era represented the first flowering of anti-intellectualism in America, as citizens adopted "the widespread belief in the superiority of inborn, intuitive, folkish wisdom over the cultivated, over-sophisticated, and self-interested knowledge of the literati and the well-to-do" (Hofstadter 1963: 154). Legislators were not immune to such anti-intellectualism. As historian Jean H. Baker (1983) shows, they were also socialized into these anti-intellectual values through shared educational experiences that promoted challenges to authority and organized dissent, values that were later reinforced by political parties.

This anti-intellectual, democratic view of knowledge informed the institutional practices of the legislature. While the legislature was by no means a bastion of measured deliberation, its structure of decision-making processes stressed debate and encouraged the exchange of competing viewpoints. Pressing issues were sent to committees for investigation. Committees gathered information (often statistical data) and submitted reports, along with their recommendations, to the entire chamber. Committees were required not only to give their recommendations but also to *demonstrate* their reasons for their conclusions. Thus, even if political machinations operated behind the scenes, in its formal procedures, the legislatures championed ideals of open debate. Arguments needed to be justified in front of the whole chamber as well as to the public through the media. The operating ideal implicit in these institutional practices bowed toward a democratic epistemology in which competing arguments were presented, debated, and then collectively decided upon. The epistemologies of Thomsonians and homeopaths, which sported flatter hierarchies in knowing, mirrored the changing conception of the public evident in the New York State legislature. Alternative medical movements did not succeed in the legislature simply because they rode a favorable cultural wave. Nor was it because they spoke the legislature's language or because they successfully framed cholera. Rather, the epistemologies tacit in their rhetoric aligned with the legislature's own understanding of knowledge.

Take for example the ideological resonance between legislators and Thomsonians—their shared commitment to common sense, anti-intellectualism, and self-sufficiency in matters of knowledge. Both viewed knowledge production through the lens of democracy and the egalitarian ideals of the Jacksonian era. The *Boston Thomsonian Manual*'s (1841, 98) complaint that "the art of healing, as it is called, has been long confided to a few interested individuals, who by degrees obtained sufficient power and influence to sway the public mind and chain posterity to whatever they might indicate for the health and lives of the people" could have easily been written by Jacksonian political reformers. Just as the people should choose and challenge their politicians, so should patients choose and challenge their doctors. In many ways, Thomsonians applied Jacksonian political logic to medicine, for their appeal to the common man, laissez-faire economic arguments, and condemnation of monopolies mimicked that of Andrew Jackson and the Democratic Party. Thomsonian rhetoric against monopolies was familiar to legislators, who often employed similar arguments in political matters, like the controversy over the Second Bank of the United States.[9]

Similarly, the sympathetic hearing given to the homeopaths by the New York State legislature stemmed from the resonance of their statistical rhetoric, and the fundamental epistemic assumptions underlying this rhetoric, with the legislature's epistemology. This elective affinity (Weber 2002 [1909])[10] between homeopaths and the legislature operated on a number of levels. First, there was a technical affinity. Both homeopaths and the legislature were engaged in gathering and producing statistical data. In a period of negative liberalism, when the government remained hesitant in acting, "one thing a government that saw itself as doing very little did do was gather statistics" (Kelman 1987, 275). Legislatures were familiar with statistical and numerical data, as they were prevalent in policy debates. Second, the homeopathic approach to knowledge production shared certain homologies with the legislature's own understanding of debate and deliberation as expressed through its institutional practices. The decision-making processes of the legislature stressed demonstration and encouraged the exchange of competing viewpoints. Contrasted with the authoritative testimony of allopathy, which attempted to convey medical knowledge through a monologue, homeopathic rhetoric embraced transparency and dialogue. Finally, the rise of statistical thinking and numeracy was tied up in the very changes in the conception of the public during this period as embodied in the legislature, which mandated that claims had to be demonstrated to the public and endorsed by it in order to be legitimate. No longer passive bystanders, the public was encouraged to participate both in the politics of the day and, for alternative medical sects, in shaping medical knowledge.

This epistemological resonance explains the success of alternative medical sects in convincing the New York State legislature to repeal the licensing laws, despite strong allopathic resistance. Thomsonians took the lead in the fight for repeal. In 1834, they gathered over thirty thousand petitions in a successful effort for the repeal of a law prohibiting unlicensed practitioners from using botanicals when treating patients (Haller 2000). Homeopaths also petitioned against licensing, and while they accumulated fewer petitions, they had the support of their more influential patients. After a decade of fierce advocacy and near misses, the legislature voted to repeal all restrictions on medical practice in 1844, placing the alternative sects on equal legal footing with allopathy.

The legislature's official justifications for the repeal reflected the affinity between the legislatures and the alternative medical epistemologies. Members of the New York legislature made two arguments for repealing the statutes, both of which demonstrated logic aligned with the democratized epistemologies of homeopaths and Thomsonians. First, the legislature respected the right and ability of citizens to draw their own conclusions about medicine. They were unwilling to infringe upon citizens' right to exercise common sense and make their own decisions. The New York Senate Select

Committee on Petitions (1844, 1), appointed with the task of assessing the antilicensing petitions, argued, "A people accustomed to govern themselves, and boasting of their intelligence, are impatient of restraint; they want no protection of their intelligence but freedom of inquiry and freedom of action." The New York Assembly Select Committee on Petitions (1842, 1) defined the Thomsonian program in terms that mirrored the Senate's vision of public discretion: Thomsonians "believe the people are capable of appreciating worth and learning in the physician, and if unembarrassed by any other circumstances will do so." Concerns about protecting the public from quackery were dismissed as unnecessary, for "light and knowledge are breaking through the barriers of darkness and superstition; education has taken deep root in the minds of an enlightened people" (New York Assembly Select Committee on Petitions 1844, 3). Believing in enlightened common sense, legislators put their trust in the people to make medical decisions for themselves. The restrictive laws were deemed illegitimate, for they contradicted the "spirit of republican institutions" (New York Assembly Select Committee on Petitions 1844, 2). Like homeopaths and Thomsonists, the legislature was willing to put its trust and faith in the common sense of the public.

Second, the legislature endorsed a view of knowledge production that stressed argumentation and debate as the means to achieve better knowledge and, in turn, better policies. The development of science, according to this view, necessitated free and open debate. The New York Senate Select Committee on Petitions (1844, 3), employing an analogy to the freedom of religion, argued,

> Science is not confined to medical societies or colleges, more than religion to councils or synods; the one is not more the providence of the legislative regulation than the other. The march of error has ever been the same; men are not content with propagating their creeds and leaving reason free to combat or embrace them—they ask the strong arm of government to decree the true religion and the true science for the sake of the souls and bodies of the people. . . . It is not the province of your committee to decide which school of medicine is the true one; it is enough to say, we believe it is the duty of the legislature to interfere with none, and to protect free inquiry.

In justifying repeal, the legislature promoted debate, recognizing no one *way* to knowledge as inherently superior to others, for to do so was to "march to error." Unlike under the politics of assent, learned men should not automatically command deference. Instead, the legislature championed the importance of debate in cultivating knowledge. Rather than discuss the *merits* of Thomsonism or homeopathy, the legislature took the occasion of licensing disputes to discuss the approach to medical knowledge generally. In this, the New York Assembly Select Committee on Petitions (1843, 7) shared a vision of knowledge similar to the Thomsonian and homeopathic view:

> And it is also clear that in the minds of your committee that such enactments operate to *restrain* rather than incite research and investigation into the hidden truths of science, by placing it in the power of one school of the profession, encircled as they now are by the strong arm of the law, to apply the epithets of quack and empiric with great force and effect to those (perhaps equally scientific as themselves) who in their investigations venture to overstep the prescribed limits of the legalized profession, and discover what to their minds is the evidence of error in the old system, and reason sufficient to induce them to propose a new and different one.

Medicine would benefit from a free market of ideas.

The importance of this epistemological resonance is underscored by the fact that the legislature supported repeal even though it remained skeptical about the validity of the Thomsonian and homeopathic systems of medicine. Indeed, many legislators were vocal in expressing their contempt toward Thomsonian and homeopathic ideas. In considering the incorporation of a homeopathic college in 1846, the Select Committee on Medical Colleges and Societies of the New York Senate (1846, 4) portrayed homeopathic thought in a decidedly negative light:

> Your committee feels that whatever may be their present convictions regarding the importance of homeopathy, as a new era in the treatment of disease, it is impossible to deny great industry to its advocates, and a furious zeal in the propagation of its doctrines. . . . And it certainly does require zeal of no common character to uphold principles *so adverse to common sense* (emphasis added).

Such negative assessments are rife in the legislative reports. Committees produced either scathing or ambivalent critiques of the alternative medical systems but supported repeal anyway. The New York State legislature may not have believed in the merits of the alternative medical systems, but it did recognize their right to take place in the debate as legitimate knowers. Without formally recognizing the legitimacy of Thomsonian or homeopathic thought, the legislature endorsed their vision of an open market of ideas, much to the chagrin of regulars: "And they [the committee] also most confidently entertain the belief, that a *discerning and enlightened people* will ever be found to award to it [a given medical sect] a generous confidence and due appreciation commensurate with its merits" (New York Assembly Select Committee on Petition 1843, 6, emphasis added). The logic of the repeals, therefore, was based on a vision of *how* medical knowledge should be produced (i.e., by open debate), not on the merits of the particular medical systems.

Regulars tried to deny the legitimacy of the alternative medical movements' claims by defining them as invalid. They denounced Thomsonians as quacks, likening them to religious fanatics and primitives who sought a return to the medicine of the Dark Ages. As the editors of the *Boston Medical and Surgical Journal* (1842, 217) argued, "Neither has the quack a right to meddle with what, to him, is as incomprehensible as a steam engine to a Hottentot." Dismissing homeopaths similarly, regulars condemned all efforts to air medical debates in public. The crux of their position was to deny the public's (and by extension, the legislatures') capability to judge medical knowledge. Their hierarchal view of knowledge held that medical knowledge could emanate legitimately only from within the ranks of regular physicians. According to regulars, "the prejudices of the public are always on the side of feeling, and never on the side of reason" (Hull 1840, 60). Medical matters were simply too complicated for the public to understand. Worthington Hooker, future president of the AMA and author of a scathing critique of homeopathy, claimed, "The science of medicine is so much a mystery to the common observer, that he cannot, as you have already seen, apply his tests to a direct examination of the physician's knowledge. He is not competent to make the estimate in this way; and if he is not aware of this, he will certainly be deceived" (Hooker 1849, 227). The denigration of the public as knowers was manifest in allopaths' use of Latin, their resorting to authoritative testimony, and their refusal to comparatively test their system against homeopaths or Thomsonians, instead dismissing them as quacks, unworthy of serious engagement. Fundamentally, regulars condemned outsiders for meddling in medicine as it contradicted their hierarchal epistemology. They rejected the legislature's ruling on medical matters, refused to endorse open debate as a valid way to attain better knowledge, and decried the pernicious effects that the repeals would have. Joseph Bates (1849, 26) expressed regulars' concerns:

> What effect has this? It has this democratic effect: It not only allows, but even holds out faltering inducements for every person to tamper with disease, and trifle with human life!! It cripples the energies of medical science, and degrades and paralyzes the influence of the profession. It throws open wide the fountains of wretchedness and crime, from which emanate those foul exhalations, that supply the polluted torrent whose invidious billows break, and dash against the time honored temples of medical science, infecting and poisoning the social horizon with their spray, far more malignant than the miasmata from the Pontine marsh.

Allopaths believed that this "democratic effect" would spell disaster for medicine.

Their arguments, however, fell on deaf ears, as allopathic appeals to paternalistic authority and their restricted notion of the community of legitimate knowers clashed with the democratic culture of the legislatures. This disconnect was not lost on regulars; the *New York Journal of Medicine* (1844a, 283) lamented that "a large portion of the public think that education and science are not necessary to qualify men for medical practice." Regular physician Dan King (1849, 371) wondered how it was possible that "the wild man of the forest, or the wilder quack in society, is deemed amply qualified" to practice medicine. For allopaths, the repeal of medical licensing laws represented nothing less than "a triumph of quackery over the Medical profession" (*New York Journal of Medicine* 1844a, 283).

CONCLUSION: KNOWING IN DEMOCRATIC SPACES

Disputes over licensing laws followed a similar pattern in other states, producing time and again the same outcome of repeal (see table 1.2).[11] Alabama and Ohio repealed their laws in 1833, hardly waiting for cholera to even clear out. From there, repeal spread throughout the thirteen states that had licensing laws on the books. In Georgia, alternative medical sects argued that licensing laws were monopolistic and ultimately detrimental to medicine ("Have not some of the most important discoveries in science been made by those in the humblest walks of life?" [Powell quoted in Haller 2000, 132]).

Table 1.2. Repeals of state medical licensing laws, post-1832

State	Licensing legislation change
Alabama	1833 Thomsonians exempted from regulations; repeal of penalties on unlicensed practice
Ohio	1833 Repeal of all penalties on unlicensed practice
Mississippi	1834 Repeal of licensing law
Massachusetts	1835 Unlicensed practitioners no longer barred from suing for fees
Georgia	1837 Repeal of sanctions on unlicensed practice; no qualification for practice other than the assumption of the name *Botanic Doctor*
Maine	1837 Repeal of licensing laws
Maryland	1838 Unlicensed practitioners no longer barred from suing for fees
South Carolina	1838 Fine and imprisonment penalties repealed for unlicensed practice; 1845 Prohibition on suing repealed
Vermont	1839 Repeal of licensing laws
Delaware	1843 Botanic, Thomsonian, and homeopathic practitioners exempted from regulation
Connecticut	1843 Repeal of licensing laws
New York	1844 All laws prohibiting persons from suing for fees for medical practice repealed; No person liable to criminal prosecution for practicing medicine without a license
New Jersey	Did not repeal until 1864

They won repeal in 1837, with the legislature eliminating all former restrictions so that they "shall cease to operate on, or have any relation to any free white person now practising, or who may hereafter, practice medicine in this State" ("An Act" March 4, 1837, 1) . In Maryland, where Thomsonians refused to be "slaves to medical aristocrats" (Rose 1838), no licenses were required after 1839.

State after state, alternative medical sects effectively transformed cholera into licensing repeals, validating their democratized medical epistemologies in the process. Thomsonians and homeopaths racked up victories, not because the legislatures necessarily agreed with their knowledge claims (and specifically those regarding cholera), but rather because they supported the *way* in which they envisioned medical knowledge. However cholera might be defined, the state legislatures recognized openness as an ideal in knowing and established an unregulated medical system that ensured the perpetuation of the debate. In 1843, the New York Assembly Select Committee on Petitions (1843, 5–6) articulated what would be the legislatures' approach to medical knowledge for the remainder of the century:

> Is it natural or reasonable to suppose, that if left free to act upon their judgments and the unbiased dictates of reason, exempt from the influences of sympathy, the people would be any more willing to entrust their health and lives in the hands of known, ignorant and unskillful pretenders to the healing art, than they would their money or property in the hands of a pretender to mechanical knowledge, without evidence of skill in or acquaintance with the pursuit in which he proposes to render service? . . . Your committee cannot conceive that such would be found to be the fact or result; nor do they see any good reason to believe that those who should be disposed to enter upon the practice of the honourable and responsible profession of physic and surgery, would suppose themselves absolved from any of the obligations or necessities of acquiring a full and perfect knowledge of the science, when placed before the public, to rise or fall entirely upon their own resources and merits. But on the contrary, that it would open a broader field of competition, operating, as in all other pursuits, to produce greater and more efficient exertion to qualify themselves to meet and successfully combat their competitors for eminence and fame.

Better medical knowledge would be obtained through competition, not control. State legislatures refused to recognize one sect over another as the legitimate source of medical knowledge; instead a "broader field of competition" won out.

The outcomes of these licensing disputes reflected a basic tension in claims to professionalization and expertise in democratic contexts—a tension that persisted throughout the epistemic contest over medical knowledge. The democratic understanding of how knowledge is best achieved (i.e., openness, transparency, and debate) clashed with the hierarchical vision of knowledge (i.e., restriction, autonomy, and noninterference) proffered by allopaths in justifying their professional privileges. The formal knowledge upon which professionals claim autonomy can be a threat to democratic decision-making (Freidson 1986). Alternative medical movements played off the tension between democracy and expertise to advance their own agendas. Without a compelling argument to grant this control over knowledge, institutions such as state legislatures were hesitant to impart it. Indeed, throughout the history of the epistemic contest, allopathy faced repeated resistance when trying to justify its claims to authority and autonomy in public institutions of the state—problems that would ultimately lead it to adopt a professionalization strategy that sought to evade state institutions.

In focusing on the rhetorical strategies of medical sects within state legislatures, this chapter offers a more nuanced way to think about the changing fortunes of allopathic medicine in the wake of the 1832 cholera epidemic. In the past, historians have often resorted to macro-cultural explanations to explain the licensing repeals, viewing the deprofessionalization of allopaths as caused by the changing cultural winds, with the spirit of Jacksonian democracy serving as the catalyst for revolts against professionalization (e.g., Marks and Beatty 1973; Rothstein 1992; Starr 1982). Such macro-cultural accounts misleadingly marshal Jacksonianism as a causal explanation for its very components. But macro-cultural shifts do not exist in some amorphous ether, hovering above social life; they are aggregations of local-level practices spanning a variety of contexts. Homeopaths and Thomsonians did not ride the wave of Jacksonian values. Rather, they actively seized the opportunities afforded by this context to transform the 1832 cholera epidemic into a successful legislative campaign and an epistemic contest. Through framing, rhetoric, and political activism, they took advantage of the Jacksonian cultural shift to achieve a particular end; in doing so, they reinforced and contributed to this shift. Without the intervention of Thomsonians and homeopaths, the legislatures would not have been compelled to act, and the licensing laws may have endured. In the end, it is actors, real breathing people, who bring about cultural change. Epistemic contests don't just happen; they are made.

Moreover, they are made in particular contexts or arenas. This chapter also calls attention to the manner in which local conditions shape the nature and outcomes of epistemic contests. Localizing the epistemic contest as such helps narrow and specify what it means when sociologists say that such and such a frame has "resonance" (Benford and Snow 2000). Macro-cultural accounts define a larger cultural context and then ex post facto deem a particular strategy as "fitting" this context. By embedding practices within institutional contexts, we can better see just *how* resonance occurs *in practice*. In this case, it was the democratizing intent of the alternative sects' epistemologies, expressed through their frames, that resonated with the burgeoning cultural and organizational commitments to democracy in the state legislatures.

The 1832 epidemic may have led to the birth of the epistemic contest over medical knowledge, but by no means did it settle it. What it did was fundamentally alter the medical market by making it more accessible to outside challengers and alternative medical movements—a situation that would endure throughout the nineteenth century. In this expanded and flattened terrain, allopathy could not simply appeal to authority; it had to either *convince* the public that its claims to authority were justified or find other ways to capture authority. As would be typical of regulars throughout the nineteenth century, they chose the latter option.

2

THE FORMATION OF THE AMA, THE CREATION OF QUACKS

THE OFFICIAL INEPTITUDE TOWARD CHOLERA, SO REDOlent in 1832, repeated itself in 1849. If the 1832 epidemic was a tragedy, the 1849 epidemic would have seemed a farce, were it not for the increase in corpses.

The nearly two decades that separated the two epidemics bore vigorous medical debate and an intensifying epistemic contest, but little insight. Answers to the most basic questions regarding cholera remained as elusive as ever. The bewilderment induced by the disease is well illustrated in a muddled description of cholera in an 1849 issue of the *Boston Medical and Surgical Journal* (1849, 123):

> Here, now, are singular facts, plainly showing the mysterious and capricious character of this dreadful disease. It appears, here, there, elsewhere, suddenly, and often giving no warning, without reference to lines of travel, regardless of natural water courses, wholly independent of the direction of the prevailing winds, and uncontrolled by the topographical character or geological formation of the districts within its general course. Spending itself where it lights first, either gently or ferociously, it disappears, and while neighboring points are standing in awe of its proximity, and daily expecting its desolating presence, it suddenly appears in altogether another region, a hundred or two miles away. And again, two or three weeks, or two or three months afterwards, while those who seemed to have escaped are still warm in the congratulations of each other, and are beginning to talk and write about the superior healthfulness of their towns, the destroyer retraces its steps, strikes their best and their worst, the strong and their feeble, alike, and carries mourning to every household.

This frustrated description of cholera could have been easily written in 1832. Cholera remained a baffling foe, its capricious nature making it impossible to pin down. Given the lack of progress in medical knowledge on cholera, it is not surprising that the 1849 epidemic followed a similar script of medical ineffectiveness, useless interventions, widespread panic, and death. Not learning from the past, the United States was doomed to repeat it.

The disease followed much the same route as it had in 1832. Once again, the Atlantic Ocean failed as a barrier. On December 1, 1848, cholera returned to New York City on a ship carrying passengers prostrate with the disease from Le Havre, France. With no quarantine station or cholera hospital established, city officials refused to let the ship dock, sentencing its passengers to a horrific stay anchored just beyond the shore. Thirty of the three hundred passengers on board died. Fortunately, the onset of frigid weather prohibited cholera from spreading beyond the harbor (Duffy 1968), despite the fact that many passengers had escaped the infected ship, reaching mainland via small boats (Rosenberg 1987b, 104).

Cholera reasserted itself in the summer, once again stirring complacent physicians and city officials into action. In May, the city established a special Medical Council of three prominent allopathic physicians, charged with containing the epidemic. They promptly declared cholera to be noncontagious (Duffy 1968). Aimed at assuaging growing panic, this announcement led to a series of misguided policies and interventions. If city officials and allopathic physicians had learned any lessons from 1832, they did not show it. The program was largely the same—clumsy sanitary and quarantine measures, polarizing debates over the nature of cholera that delayed every proposed action, and the continued use of heroic treatments to the detriment of patients. In a testament to such stagnation, the council reprinted the same 1832 broadside of recommendations to the public with only a few "verbal changes" (Rosenberg 1987b, 109). Seventeen years after the original outbreak of cholera, officials had little more to offer than "Be Temperate in Eating and Drinking!"

As the epidemic intensified, hostility toward health officials grew. The Medical Council's policies were met with widespread skepticism and, occasionally, outright revolt. When the council took over and transformed four city schools into temporary cholera hospitals, it incited ire from the press and local communities in the process. People avoided the new hospitals at all costs; to be sent to them was seen as a death sentence. One of the council's rare innovations was a massive effort to remove pigs and other livestock from the streets. As of 1842, there were a recorded ten thousand hogs roaming the

CHOLERA!

Published by order of the Sanatory Committee, under the sanction of the Medical Counsel.

BE TEMPERATE IN EATING & DRINKING!

Avoid Raw Vegetables and Unripe Fruit !.

Abstain from COLD WATER, when heated, and above all from *Ardent Spirits*, and if habit have rendered them indispensable, take much less than usual.

SLEEP AND CLOTHE WARM !

DO NOT SLEEP OR SIT IN A DRAUGHT OF AIR,

Avoid getting Wet !

Attend immediately to all disorders of the Bowels.

TAKE NO MEDICINE WITHOUT ADVICE.

Medicine and Medical Advice can be had by the poor, at all hours of the day and night, by applying at the Station House in each Ward.

CALEB S. WOODHULL, *Mayor.*
JAMES KELLY, *Chairman of Sanatory Committee.*

1849 cholera broadside, distributed by the New York Medical Council, reprinted almost verbatim from the 1832 broadside. The Granger Collection, New York.

city. Because the council believed that there was some connection between cholera and filth, it decided to remove the pigs as part of its effort to clean up the city. Undoubtedly an eyesore—Charles Dickens described them as "having, for the most part, scanty, brown backs, like the lids of old horse-hair trunks; spotted with unwholesome black botches . . . long, gaunt legs, too, and such peaked snouts, that if one of them could be persuaded to sit for his profile, nobody would recognize it for a pig's likeness" (Dickens 2000, 97)—the pigs performed an essential service for the city. By removing the pigs, officials removed one of the most important scavengers from the streets at the precise time they needed them, increasing the city's filth and facilitating cholera's spread (Duffy 1968; Rosenberg 1987b). Moreover, these removal efforts were met with riots, as the hogs were essential to the livelihood of the poor.

These poorly conceived plans lacked physician support, as doctors and city officials continued their mutual disdain. Physicians, fearing the loss of clientele to the council, were reluctant to report cholera cases or turn over their patients to cholera hospitals. This inaction was met with a sharp rebuke from the council. The bickering and lack of coordination fueled the public's misgivings about authorities' ability to combat cholera. By June, many had fled the city, and by July, the height of the epidemic, business had halted to a standstill.

For all its tribulations, New York City's experience with the 1849 epidemic was comparatively mild by national standards (it had claimed over 5,017 victims, 1 percent of the city's 515,000 inhabitants [Alcabes 2009]). But cholera traveled more widely in 1849 than it had in 1832. After arriving in the port of New Orleans on December 6, cholera rapidly spread inland throughout the South. Terror accompanied the disease wherever it went. Mark Twain, chronicler of steamboat life along the Mississippi River, recalled the hysteria: "The people along the Mississippi were paralyzed with fear. Those who could run away, did it. And many died of fright in the flight. Fright killed three persons where the cholera killed one" (Twain 2010, 352). As for those who couldn't flee, they "kept themselves drenched with cholera preventives" (Twain 2010, 352). Perhaps fortified with the fearlessness of youth—Twain, a teenager at the time, remained calm. Rather than stomach the gruesome cholera preventive his mother prescribed, he poured it down the floorboards, to "good result" as "no cholera occurred down below" (Twain 2010, 352).

Twain's experience with the disease emphasized the link between cholera and economic development. Cholera followed improvements in transportation, most notably the extension of railroad and steamship transportation. The Gold Rush of 1849 brought the disease west with disastrous results. "It was in the infant cities of the West," according to Charles Rosenberg (1987b, 115), "with no adequate water supply, primitive sanitation, and crowded with a transient population, that the disease was most severe." Western cities like St. Louis, which lost 10 percent of its inhabitants, were devastated. Even isolated towns did not escape cholera's reach. Based on incomplete statistical data, the AMA later estimated that cholera had claimed over thirty thousand victims (Newman 1856), one of which was former president James K. Polk.

Local officials scrambled to do anything to halt cholera's spread. In Pittsburgh, where "great fear came suddenly upon the people, and the excitement was unbounded," officials adopted a two-pronged attack; they removed "damaged vegetables" from the market and burned tar throughout the city ("The Cholera at Pittsburgh" 1854, 4166). It was in Cincinnati, however, that cholera reached its absurd apogee. There the city council spent $3,000 to fund a project of burning Youghiogheny coal at street crossings in an effort to alter the poisonous atmosphere. The *Christian Advocate Journal* ("A City in Mourning" 1849, 119) dutifully reported the program's inevitable failure: "The coal was duly fixed and fired, and had mixed with it large quantities of sulphur and tar, but neither the coal nor the tar had any effect in scaring off the disease." Compounding this inevitable failure, the Cincinnati Board of Health became a lightning rod for medical disputes; as one observer noted, homeopaths, eclectics, and allopaths were "each jockeying for position on the Board of Health, at the city hospital, and for favor with the public" (Chambers 1938, 218). Rendered impotent by these sectarian disputes, the board was summarily dismissed midway through the epidemic. *All* physicians were removed from the board, replaced by a lawyer, an editor, a liquor dealer, a preacher, and a mechanic (Rosenberg 1987b, 118). The city's curious cholera experiences were not restricted to the dubious exploits of the board. In a single day, the Cincinnati press reported on a cholera patient who rose from the grave—an event mildly referred to in the title as a "cholera incident" ("Cholera Incident" 1849, 124)—and an entire family of six who succumbed to the disease—the father, dying not from the disease directly, but from grief after watching his wife and four children succumb ("A Sad Story—Effect of the Will," 1849, 124).[1] Stories like these proliferated throughout the country, spawning many cholera myths that intensified anxiety.

As before, physicians could offer little in the way of a coherent explanation for cholera, let alone effective treatments. Homeopaths argued with allopathic physicians, rationalists with empiricists, contagionists with non-contagionists. Indeed, doctors were still disputing whether cholera was in fact a unique disease in its own right. This prompted one allopathic doctor to beg his fellow physicians "to discard the name of cholera in his treatment of the disease, and when called to the bed-side of the patient come untrammeled by any specious reasoning" (Page 1849, 3). Doctors deployed a variety of exotic and none too effective treatments, which, according to one allopathic physician, were "unwarrantably conjectural, experimentative, and often entirely irrational" (Dick 1849, 3).

Wondering "who shall decide when doctors disagree" (Lynde 1848, 2), many Americans once again sought refuge in religion. On August 3, President Zachary Taylor called for a national day of fasting to purify the country's soul and to stifle the spread of cholera. Clergy and moral reformers drew on the notion of predisposing causes to admonish their flocks to change their ways (Rosenberg 1987b). To them, cholera was a killer of the sinful. "Hecatombs of the licentious have been sacrificed, and multitudes of inebriates have fallen easy prey to the ravager," noted the *New York Evangelist*. "Intemperance in eating or in drinking—excessive indulgence of any kind—and all illicit indulgences are calculated to invite the disease and increase its violence" ("Best Preparation for the Cholera" 1849, 94). Responding to this outbreak of piety, indulgers and inebriates did not sit idly by as their lifestyle was condemned on medical and moral grounds. Brewers in Baltimore claimed that malt liquor made one immune to cholera ("Malt Liquors and the Cholera" 1849, 125), even mustering statistical data in an effort to convert skeptical teetotalers.

As before, cholera lost much of its vigor once summer ended. Cases were documented intermittently every year between 1849 and 1854, even reaching epidemic proportions in 1854. For five years, the country was on edge, but after 1854, cholera disappeared again.

Still its effects on medicine endured.

ALLOPATHY RESPONDS

The 1849 cholera epidemic unfolded within a different professional terrain, as the epistemic contest had evolved. The program of legislative reforms carried out by homeopaths and Thomsonians after 1832 had been incredibly successful. With licensing protections revoked in state after state, the medical market opened, fueling the epistemic contest and encouraging more vigorous competition between medical sects.

The makeup of the competition had also changed. Subsequent to the successful campaign to get licensing laws repealed, Thomsonism fragmented, succumbing to infighting after the death of Samuel Thomson. In many ways, the repeal of the New York State law, in which the Thomsonians played such a key role, was the final salvo of the movement. The schism revolved around a number of issues, including fights over succession, financial squabbling, and competing proprietary claims. While Thomson himself had long served as a lightning rod for these very issues, the strength of his personality as well as his ability to centrally control the movement through the distribution of its patented system kept the problems in check (Berman and Flannery 2001; Haller 2000). After his death, however, the suppressed problems erupted, nearly destroying the movement. The seeds of discontent coalesced around debates over whether Thomsonians should establish a formal education system (Haller 2000). Reformers called New Light Thomsonians sought to establish medical schools and formal training, a plan anathema to Thomson's original intent of making every person his own physician (Haller 2000). Unable to resolve the tensions between the New Light Thomsonians and the more traditional Thomsonians, the movement split. Some members reorganized into a broader, loosely defined sect known as Eclecticism, named such because of its eclectic adoption of ideas from a variety of medical sects. Eclecticism would endure until the twentieth century, but its ecumenical pragmatism blunted the original democratic critique of Thomsonism. The movement never again mustered the same type of epistemological challenge to allopathy but instead was content to follow homeopathy's lead.

Homeopathy stepped into the breach, establishing itself as *the* main challenger to allopathy—a position it would hold into the twentieth century. Homeopaths sought to democratize medicine in a more tempered way than Thomsonians. To regulars' opacity and elitism, homeopaths invited the public to assess competing knowledge claims. They also offered a sophisticated system of medical knowledge and an articulated epistemological program that claimed the scientific mantle through an appeal to empiricism. And regulars' recycled—and ineffective—response to the 1849 epidemic emboldened the homeopathic challenge. Homeopathy's popularity grew, for the second epidemic confirmed for many what was hinted at during the first—in opposition to allopathy's heroic treatments, homeopaths offered milder treatments which may not have done much positive good but certainly did not have the same type of negative effects on patients as allopathic therapeutics (Coulter 1973; Haller 2005; Kaufman 1988; Rothstein 1992; Warner 1997). Adept at using statistical rhetoric to support this claim, homeopathic ranks swelled (Coulter 1973), and many allopathic physicians converted to homeopathy (Haller 2005; Kaufman 1988). On the ascent, homeopaths began to challenge allopathy on a number of different fronts, demanding inclusion in all aspects of medicine.

Cognizant of the changing dynamics of the epistemic contest, allopathic physicians turned their focus to the homeopathic challenge. This chapter chronicles regulars' response to the ascendant homeopaths. Even prior to 1849, reformers within allopathy sought to reformulate the epistemological foundation of allopathy and establish an organizational infrastructure to deal with the vitalized "quackery." As alternative medical movements shifted medical debates onto epistemological grounds, allopathic reformers felt the pressure to respond in kind. Their epistemological response was to discard rationalism in an effort to establish a new, more coherent epistemological foundation for its knowledge. They promoted a program of radical empiricism inspired by the Paris School of medicine, which sought to ground medical knowledge, not in abstract speculative systems, but in sensory, bedside observations. In doing so, reformers hoped to tap into the democratic zeitgeist that alternative medical movements had captured so effectively.

However, in rejecting the universalizing impulse as retrograde rationalism, allopathic physicians were left with a dearth of epistemological standards to adjudicate between competing allopathic claims and, more troublesome, to dispense with competing homeopathic claims. To resolve this "problem of adjudication," reformers adopted an *organizational* strategy that would have long-lasting effects; in 1847, a group of elite allopathic physicians established the American Medical Association (AMA). The second section of this chapter examines the formation and early practices of the AMA as an organizational response to the epistemological problem of adjudication. The early AMA brought a measure of cohesion to allopathy and established itself as *the* major organizational player for the future of the epistemic contest. In turn, it affected the content of allopathic knowledge produced on cholera by creating a community of knowers insulated from outside influence that rejected methodologies associated with homeopaths. While this strategy had mixed *professional* effects, it would shape the contours of the epistemic contest over cholera for the remainder of the century.

ORGANIZATIONAL PRACTICES AND EPISTEMIC CONTESTS

Historians have longed dismissed the AMA's early activities as ineffective and inconsequential (e.g., Burrow 1963; Duffy 1993; Starr 1982; Stevens 1971). The organization initially outlined three goals—educating the public on medical issues, reforming allopathic medical education, and combating "quackery" in all its guises through legislative efforts. Yet even generous assessments of the early activities of the AMA acknowledge that it failed to make much progress along any of these lines (see Fishbein 1947; Porter 1998). Indeed, the early period of the AMA, from roughly 1840 to 1880, witnessed the declining status of the profession (Numbers 1988).

While correct in pointing out the AMA's inability to achieve its specific professional goals during its first fifty years, these accounts overlook an important epistemological function of the early AMA. This oversight is not just a product of damning historical evidence; it stems from the way in which previous research conceptualizes professionalization as an abstract process, one that contains underlying teleological assumptions that blind researchers to the role of historical contingencies in the development of professions (Freidson 1986). Because the early AMA failed to move allopathy closer to the ultimate goal of the monopolization of work, it is dismissed as unsuccessful and even irrelevant.

By embedding the formation of the AMA and its initial practices within the context of the ongoing epistemic contest over medical knowledge, this chapter offers a corrective reinterpretation of the early AMA. Rather than reading this early history through an abstract, ideal-typical conceptualization of professionalization, I situate the founding of the AMA within the context of the specific problems facing allopathy at the time. These were professional problems, but not generically so; they were inextricably intertwined in the epistemic contest over medical knowledge. In a period of epistemological flux and intense competition between competing medical sects, allopathy faced a problem of adjudication. How could true medical knowledge be recognized and legitimated over false belief in an intellectual environment devoid of shared epistemological assumptions? How could regulars distinguish their knowledge from homeopaths in such an environment?

What the early AMA provided was an organizational criterion that served as a proxy for epistemic standards to adjudicate knowledge claims. Organizations play an integral role in shaping the epistemological landscape, defining the "schemas of plausibility" (Shapin 1994, 22) and outlining the nature of epistemic legitimation (Biagioli 1994). Consequently, organizational formation can be an important epistemic practice. As noted in the previous chapter, organizations participate in sense-making (Weick 1979) as epistemic settings (Vaughan 1999) that establish the parameters of acceptable knowledge practice. However, because intellectual and epistemic authority is often viewed primarily as a cultural product, these epistemological functions tend to go unnoticed in much of the research on knowledge struggles as researchers tend to appeal solely to cultural factors to account for the allocation of intellectual authority. While organizational practices are shown to be integral in the *production* of science (e.g., laboratory studies show how organizational arrangements shape facts produced in the lab), the interaction between cultural and organizational practices has been undertheorized in the sociology of science "downstream," which focuses primarily on cultural practices like boundary work (Gieryn 1983, 1999), rhetorical repertoires (Gilbert and Mulkay 1984), public performances (Hilgartner 2000), and framing to account for the resolution of public debates over science and the authority of science.[2] Analyses of cultural strategies must be complemented by analyses of organizational practices, or the ways in which actors draw upon and deploy organizations to construct boundaries and capture credibility.

This is especially pertinent when examining epistemic contests, whose distinctiveness lies precisely in the absence of shared cultural valuations of knowledge. Epistemic contests are not waged by cultural means only. When the very parameters and terms of cultural debates are up for contention, there is an increased tendency for actors to speak past one another from incommensurable epistemological positions. Such incommensurability tends to undermine the efficacy of cultural arguments, spurring actors to adopt a more diverse set of strategies. In turn, organizational strategies play a more crucial role in actors' repertoires in epistemic contests than they do in typical knowledge disputes. Only by examining the *interaction* between cultural and organizational practices can the trajectory of epistemic contests be explained.

In this chapter, I identify and explore the epistemological function of organizational practices, specifically the role of organization formation in waging epistemic contests. Whereas the previous chapter demonstrated the importance of organizational *context* on the trajectory of epistemic contests, in this chapter, organizational *practices* are shown to be intricately involved in epistemic contests, shaping their trajectory. Organizations are no longer banished to the background as mere settings for cultural disputes over knowledge. In this case, organizational formation (i.e., the establishment of the AMA) emerges as an important epistemological strategy in itself. If we think of epistemological commitments in terms of "dwelling in" an intellectual system (Polanyi 1958), then organizations serve as the formal dwellings that shape the epistemological terrain for actors. Organizations are an important resource in epistemic contests as they legitimate particular epistemological positions by configuring and institutionalizing communities of knowers and marshaling resources to promote the production of knowledge along certain epistemological lines. Given these functions, actors attempt to harness the power of organizations to promote their epistemological agendas and to alter the epistemological terrain through organizational practices. This is precisely what regulars attempted to do with the AMA.

IN PARIS WE TRUST

After 1849, the epistemological fragmentation within allopathy came to a head. The repeal of the licensing laws meant that allopathic physicians could no longer hide behind legal statutes when challenged by outsiders. They had been decertified. And in the open medical market, with competing medical sects offering alternative epistemologies, inchoate epistemological commitments no longer sufficed. Finding the status quo of internal divisions over the old rationalism and crude proto-empiricism untenable in such an environment, elite reformers within allopathy searched for a new epistemic foundation for their medical knowledge. They found it in Paris.

To regain their status, allopaths needed a coherent epistemological account of what it was they were doing, and in the 1840s, a group of reformers embraced a vision adapted from the Paris School. Reformers' motivations were multiple (Warner 1998, 12). Many came to reforms rather unintentionally. As young doctors trying to gain a leg up in the overcrowded antebellum medical market, some traveled to Paris, the world's center of medical innovation, for additional education, only to become socialized into a new vision of medicine (Warner 1998). Some were dissatisfied with the heroic practices and intellectual stagnation of the previous generation. For these doctors, the very term "system" had become an epithet of derision (Warner 1997, 50), and they gravitated to the next new thing. And others, concerned about the waning status of allopathy, consciously sought collective uplift through a reformulation of medicine along "scientific" lines. Whatever their original motivation, the driving question for all reformers became, how could allopathy move beyond rationalism and capture public acclaim in the context of Jacksonian democracy?

Between 1820 and 1861, nearly seven hundred doctors traveled to Paris to supplement their education, forging such a strong "living link between the Parisian medical world" and U.S. allopathy (Jones 1970, 144) that this period is often called the "French period" of American medicine (Ackerknecht

1967). Initially it was not the French ideas that drew Americans to Paris; it was the institutional infrastructure of French medicine, which afforded students and doctors with access to a large cache of patients, an opportunity unavailable in the United States (Warner 1998). State-run hospitals and clinics provided access not only to a large patient population but to cadavers for autopsy as well. Additionally, the laxer sexual mores allowed physicians to gain experience working with female patients. For the elite of the U.S. medical profession, precisely the men who would become the leaders in medical societies, schools, and journals, studying in Paris became a rite of passage that shaped their professional identities through the mid-1800s. At least a third of this number would end up teaching at American medical schools (Jones 1970). Given the influence of this group, the Paris School as an ideal became the "most powerful source of change in antebellum American medicine" (Warner 1998, 4).

Drawn to Paris's practical educational opportunities, these doctors came to embrace its medical epistemology: "Paris is confessedly the seat of medical knowledge, and undoubtedly comprehends within its precincts the greatest body of men learned and eminent in every department of science, and especially in that of the healing art" (Yates 1832, 1–2). At its most basic, the Paris School embraced empiricism as the foundation for medical thought, as it sought to erect a body of knowledge built from careful empirical observations. It stressed systematic physical examinations of patients, "an anatomo-clinical paradigm rooted in the systematic correlation of signs and symptoms observed at the bedside with lesions found in the organs at autopsy" (Warner 1998, 4). Employing new techniques like auscultation, the Paris School championed an empirical approach to medicine in which patients were examined thoroughly both externally through assisted sensory observation and internally through autopsy. By extending the "medical gaze," the Paris School sought to penetrate superficial symptoms to get at the root cause of disease (Foucault 1994). This empiricism was embodied in the stethoscope (Warner 1998, 137), the autopsy, and the central role of the clinic in medical training (Foucault 1994). The new tools from Paris would refine observation to such a point that the need for speculative systems would be eliminated: "Percussion and auscultation; the various forms of specula, such as the ophthalmoscope, the microscope, the endoscope, and more lately the laryngoscope, have all furnished the most useful results, and added very materially to our knowledge of diseases by dissipating those false notions of physiology and pathology, which were founded upon mere theory and speculation" (Quackenbush 1869, 4–5).

The Americans who studied in Paris returned to the United States with a firm commitment to the Paris School, viewing empiricism as the way to address the professional crisis facing them at home. Through their letters, translations of French works, and their advocacy of educational reforms, these doctors transmitted the Paris School to the United States (Jones 1970). But the Paris School that these Americans championed assumed a different identity in the U.S. context. Transmission is a selective process, not one of passive diffusion. Allopathic reformers saw the new epistemology as the antidote to the rationalism that precipitated their dwindling prestige. But they did not merely transplant the Paris School to the United States; they selectively embraced particular aspects of it, adapting it to fit their professional needs (Warner 1998). In other words, they transformed the ideals of the Paris School to address the epistemic contest with homeopaths.

In essence, for Americans, the Paris School came to represent an effort to excise all philosophical speculation from medicine so as to transform it into a science built on observations and facts. In setting the Paris School in opposition to rationalism, American reformers transformed the school's systematic empiricism, rooted in an extensive organizational infrastructure and buttressed by numerical methods, into a *radical empiricism* based on observations gleaned only through sensory input—a sensual empiricism that sought facts through sight, touch, and sound.[3] This epistemology was more radical than that of the Paris School as it rejected the validity of abstracting general theories from local observation. As such, the Paris School was reformulated in negative terms as an antisystem that sought to shear medicine of all theoretical scaffolding and deal "only with positive tangible facts" (Bissell 1864,18). Less a formal methodology, the Paris School became equated with a general orientation toward medical knowledge characterized by an "allegiance to empirical fact, to knowledge gained and verified by direct observation and analysis of nature, coupled with the opposition to rationalism, hypothesis and speculative systems of pathology and therapeutics" (Warner 1998, 8). This commitment to facts, it was hoped, would save the profession plagued by theoretical speculation divorced from reality (e.g., Bissell 1864; Quackenbush 1869).

Radical empiricism was explicitly set against Enlightenment rationalism and displayed an animus against all speculation. It attempted to purge the knower of all theoretical speculation so as to obtain more objective, fact-based knowledge. Medicine needed to become an empirical pursuit of truth based on facts, uncovered through the experienced, intelligent use of the senses. The model knower was the individual physician who diligently and thoroughly observed the particular patient. The locus of the production of knowledge was at the bedside, where the doctor interacted with and observed the illness. And medical knowledge would progress through the incremental accrual of discrete facts, grounded in tangible (visible, audible, olfactory) observations rather than flighty bouts of speculation. The metaphor driving this empirical vision was one of accumulation: "The science of medicine has been aptly likened to an ant-hill, in its slow but steady growth, no one individual adding but a mite to the mass of facts which compose this hill of science" (Jones 1861, 10). As such, reformers policed all medical knowledge for theoretical musings, deriding—and dismissing—that which appealed to generalization and speculation as illegitimate. What was not an observable fact was not allowed. While such atheoretical observation is impossible, set within the context of the epistemic contest over medical knowledge, its appeal to regular reformers was that it seemed a remedy to the disastrous professional consequences they believed followed from rationalism. In addition, it was attractive because it built upon the nascent empiricism already existing within allopathy. And because empiricism was rooted in sensory observation, making knowledge ostensibly available to everyone, it appeared more democratic, allowing allopaths to claim some democratic bona fides.

But the animus toward generalization and theorizing was often taken to extremes. In 1844, Elisha Bartlett, an influential reformer and committed advocate of radical empiricism, gave the fullest account of the new epistemology in *An Essay on the Philosophy of Medical Science* (Warner 1998, 175). Bartlett denied the legitimacy of *any* hypothesis or theoretical speculation that might bias observation; medical knowledge "*is in the facts and their relationships, classified and arranged, and in nothing else*" (Bartlett 1844, 7 emphasis in the original). Hypotheses and theories "shut up, or obscured, or perverted" the senses (Bartlett 1844, 218). Speculation corrupted observation, for "one of the first and most inevitable effects of a belief in any a priori system of medicine is an *utter disqualification of the mind for correct and trustworthy observation*" (Bartlett 1844, 206 emphasis in the original). A nomadic professor who taught at a number of medical schools in different states, Bartlett did his best to socialize the next generation of doctors into this empirical ethos. For this generation of students, medical knowledge would no longer be achieved by armchair physicians who constructed elaborate rational systems; it would be gained by staying close to sensory observations at the bedside and the uncovering of simple facts.

Strategically, radical empiricism had three benefits for allopaths in regards to the epistemic contest. First, it represented a return to the Hippocratic tradition of empirical, bedside observation upon which medicine should be built—a tradition that was lost during the age of rationalism and speculative systems. The flip side of this positive embrace of empiricism was a rejection of rational systems, allowing allopathy to shed the most obvious target of its critics (Warner 1998).

Second, radical empiricism allowed regulars to turn the rationalism epithets back at their opponents. In this way, they attempted to one-up homeopaths in their commitment to empiricism. Like homeopaths, allopathic physicians sought to claim the mantle of scientific medicine via empiricism, albeit through a different understanding of what empiricism meant. Allopathic *radical* empiricism differed significantly from homeopathic empiricism; homeopaths were not as averse to generalizing and theorizing across contexts as allopaths were. This difference stemmed from the origins of the two systems. Because homeopathic empiricism did not arise in opposition to theorization, homeopaths expected facts to accrete into a coherent body of medical knowledge *with* the aid of theory; facts were a means to the end of universal laws, like the law of similars. Allopathy's aversion to anything that reeked of rationalism, on the other hand, precluded the search for underlying rules to make sense of discrete observations. Facts *were* the end. This radical positioning allowed allopaths to portray homeopaths as rationalist because they did not go far enough in their rejection of theorization. They framed homeopaths as backward, focusing on the law of similars and the notion of the vital force as signs of their theoretical excess, rather than their empirical strivings through provings and statistics.

Finally, the commitment to observable facts implied some transparency in knowledge that resonated with the democratic ethos of the time. The success of Thomsonism and homeopathy in adopting democratic rhetoric to serve their ends was not lost on allopathic reformers. They sought to exploit it for themselves, albeit in a more conservative way; their empiricism would be accessible to outside observers but still require expertise to be done correctly. This was not to be naïve empiricism, but one undertaken only by elite observers. Still, the fact that it posited an understanding that medical insight was more accessible made it an improvement, democratically speaking, over the blatant elitism of rationalism.

Ultimately, reformers hoped that radical empiricism would yield accurate medical knowledge on diseases like cholera by stemming the impetus toward inaccurate speculation (Warner 1998). They championed a new identity for the American physician: "The *fact-hunter*, as he has been sneeringly called, provided he be also a *fact-finder*, and a *fact-analyzer*, is the only true contributor to the advancement and improvement of medical science," announced Bartlett (1844, 220). This new identity required nothing short of the adoption of an entirely new epistemology.

Proliferating Accounts of Cholera and the Problem of Adjudication

Despite the great hope reformers placed in radical empiricism, it was accompanied by its own set of problems. In Paris, empiricism was buttressed by a strong institutional infrastructure, a profession with firm support from the state, an organized community of medical elites, and a coherent program of medical education that made new medical tools widely available. The situation in the United States was much different. Allopathic physicians did not have government support and were left to their own devices to carry out their reforms. Moreover, American medicine lacked any sort of infrastructure that could bring coherence to a new epistemology reticent to theorizing across contexts. Hospitals remained on the periphery of medical practice (Rosner 1982). Medical education, lacking access to large patient populations, remained didactic, and standards remained very low (Ludmerer 1985; Markowitz and Rosner 1973; Starr 1982). Education was so limited that new doctors received little training in the science of observation. As such, the adoption of the microscope, stethoscope, and other tools of the Paris School was hindered. For example, use of the stethoscope was confined to only the most elite doctors after the Civil War (Rosenberg 1987a, 91). The senses were not being channeled into devices that could be standardized across contexts.

Without these important institutional means to organize the production of knowledge, the embrace of radical empiricism presented a problem: allopathic physicians remained committed to the centrality of sensory observation in developing a truly scientific medicine but lacked the institutions to provide coherence to this endeavor, to aggregate local observations into a generalizable knowledge. Radical empiricism, therefore, devolved into proliferating claims made by individual doctors according to their own observations. A "fact" would be observed and reported, but what if other individual doctors did not make the same observation? There was no adequate method by which to weigh one individual observation against another. There was no way to know if doctors were on the same page, or even seeing the same things. In other words, radical empiricism suffered from a problem of adjudication. How were allopathic doctors to decide when observations disagreed? The rejection of anything that reeked of rationalism precluded allopathic physicians from developing underlying rules and laws to bear on discrete observations. The animus against speculation and generalization bound medical knowledge to its local context and hindered efforts to develop a general body of medical knowledge.

Quantification could have offered a solution to the problem of adjudication. By eliminating extraneous information, numbers simplify and standardize data, making it liquid, comparable, and mobile (Carruthers and Espeland 1991; Porter 1994). As such, quantification might have offered regulars a natural escape from the local and the particular. After all, the Paris School, under the influence of Pierre Charles Alexandre Louis, promoted statistical reasoning and the "numerical method" (Ackerknecht 1967; Matthews 1995). Louis collected numerical data on patients in hospitals so as to compare treatments and differentiate diagnoses. The numerical method never caught on in the United States, despite some influential supporters like Bartlett and the prolific author and Harvard- educated physician Austin Flint. In part, this resistance reflected the institutional reality of U.S. medicine; without large, state-run hospitals it was technically difficult to collect and aggregate numerical data. In part, it reflected the transformation of the Paris School into radical empiricism in the American context. Radical empiricism stressed local observation; data that was abstracted from local-level sensory observation, or was not focused on the particular case, was foreign (or at least problematic) to its analytical orientation. Numbers could not be seen, heard, or touched. And like many nineteenth-century medical thinkers (Hacking 1990), allopathic physicians questioned whether information on collectives had relevance for the treatment of individual patients. For example, the *New York Journal of Medicine* (1844b, 327), while admitting the importance of statistics in other fields of inquiry, stated, "[statistical] laws present nothing individual, their application to individuals is only within certain limits." In doing so, they raised the common criticism that aggregate information bore little relevance to treating individual patients—a critique that Louis himself faced in France (Matthews 1995). Allopathic physicians rejected statistics for the same reason that universal laws were difficult to come by under radical empiricism. Both dealt in abstract aggregation, whereas the new epistemology was oriented toward the particular and individual.

But most important, the rejection of statistics reflected an ideological opposition. Regulars had come to associate statistics and quantification with homeopaths and, in turn, rejected it out of hand. They dismissed homeopathic statistics as rhetoric, presented in "the advertising style of quackery" (Hooker 1852, 109) so as to dupe the public. They questioned the trustworthiness of the data that underlay homeopathic statistical claims: "The value of statistics, and especially when they relate to therapeutics, depends upon the principles on which they are collected, and the mental and moral character of him who collects them. It is often said that 'figures cannot lie;' but the annals both of quackery and of medicine show, that false statements can be made as easily in figures as they can be in words" (Hooker 1852, 107–108). Because of this outright rejection of statistical reasoning, allopaths lagged far behind their European peers in the collection and analysis of vital statistics (Duffy 1990; Haller 1981; Meckel 1998).

Therefore, while allopaths had shed speculative systems that they believed undermined their professional claims, radical empiricism unintentionally produced similar fragmentation and unintelligibility toward cholera, as it prevented medical knowledge from traversing local contexts and the idiosyncrasies of the individual. It did little to improve regulars' understanding of the disease, providing only a confused mass of disjointed observations. In 1849, as in 1832, there were a number of contradictory accounts of cholera without any standards to assess and compare them. In 1832, the problem of adjudication grew out of the incommensurability of competing rationalist systems and the underdeveloped epistemological account of allopathic knowledge; in 1849, it was created by the reluctance of those committed to radical empiricism to engage in any sort of generalization. Allopathic knowledge of cholera in 1849 had changed only marginally since 1832. Allopathic physicians could recognize cholera at the bedside and were able to diagnose it consistently,[4] but they still lacked any effective therapies or any effective sanitary techniques to limit its spread. Suggestions were simply rehashed from the 1832 epidemic and included everything from wearing wool to fleeing to the country. In fact, the resistance to theorization and the fragmentation that ensued led some to adopt a posture of extreme skepticism toward all therapeutic interventions, a position of "therapeutic nihilism" (Starr 1976). And while all agreed that cholera preyed on certain predispositions, both at the local and individual levels, with no way to compare the relative influence of these different predispositions, which ranged from fear of cholera to dampness in cellars, the list of predispositions multiplied into a hodgepodge of empirical observations and basic common sense, hardly a scientific achievement.

But more than its therapeutic and preventative failures, it was allopathy's inability to provide an adequate *account* of cholera that proved most damaging during the epidemic. While there was now near-universal acceptance that cholera represented a specific disease with its own identity (Rosenberg 1987b, 149), its etiology continued to be a most vexing question. By 1849, only the rare doctor still subscribed to a theological or supernatural causal argument (Rosenberg 1987b). But the materialist explanations remained as numerous as in 1832, with additional theories emanating from the voguish sciences of chemistry, microscopy, and pathological anatomy (Richmond 1947). Cholera was described variously as contagious, noncontagious, or contingently contagious, caused by miasmic gases, consumption of alcohol, atmospheric changes, "a motivating agent," or corrupted vegetables. Juxtaposed to the tidy statistical ratio of cholera offered by homeopathy, this internal confusion did not compare favorably.

Although the terms of the debates were unstable and ill-defined (Richmond 1947), in general allopathic physicians fell into two broad camps—contagionists versus noncontagionists—with a number of theories competing for prominence within each. Contagionists were the minority. They believed that cholera spread via infectious people and drew on the disease's movement—and its tendency to cluster—as evidence. This account was somewhat undermined by the experiential observation that medical professionals attending to cholera patients rarely succumbed to the disease. Still, contagionists "assumed cholera to be the result of some specific poison" even though "nothing demonstrable is known concerning the nature of the cholera poison. All that has hitherto been advanced in this direction is pure hypothesis" (Metcalf 1869, 2). This mysterious poison was described variously as "animalculae" (*New York Journal of Medicine* 1849b), a "morbific agent" (Macneven 1849, 195), a "vegetable fungus" (Dickson 1849, 13), and "cryptogamic" (Seymour 1857, 188).

Most regulars remained skeptical of these early germ, or animacular, theories. The bulk of physicians subscribed to some sort of noncontagious account of cholera, where there was more intellectual energy. Noncontagionists focused on miasmas (poisonous emanations from the soil or filth), atmospheric causes (e.g., humidity, static electricity in the air, etc.), zymotic causes, or newer theories of fermentation inspired by chemistry. Atmospheric theories saw cholera as a form of "meteratorious" epidemic, in which the atmosphere created a poisonous condition. Fermentation theories argued that cholera originated in a specific poison which only gained lethality when "fermented" by favorable conditions. Many fermentationists offered some sort of "zymotic" causal account by which decaying organic matter released toxins into the atmosphere (Eyler 1973). However, these theories, like the contagionist theories, were plagued by ambiguities and inconsistencies. As such, noncontagiousness theories were not immune to critique. For example, atmospheric theories had trouble explaining the obvious fact that cholera traveled. Pouncing on this inconvenient fact, a doctor of the contagionist persuasion derided the atmospheric theory as merely a scapegoat used to mask doctors' ignorance (Rosenberg 1987b, 147).

Given these basic disagreements as to the nature of cholera, a coherent picture of the disease failed to develop from the copious studies allopaths undertook. The New York Academy of Medicine (NYAM) established a special committee to conduct an extensive analysis on the contagion question. Their conclusion? It was "premature and inexpedient for this Academy to pronounce at the present time any positive opinion in regard to the contagious or non-contagious nature of Cholera" (quoted in Van Ingen 1949, 39). Compounding things was the fact that there was "a want of uniformity in the mode of making reports, which obscures, or even renders inaccessible the truth" (AMA 1850, 107). The problem of adjudication, which underwrote every debate on cholera, led to the inconsistent presentation of observations. Furthermore, neither animacular nor atmospheric theories even satisfied the criteria of sensory observation (i.e., they were not readily observable through the senses) and thus remained problematic for a profession committed to radical empiricism. Etiological candidates and studies proliferated, but there was no way to weed out the good versus the bad, the true from the false.

Regulars' mood soured into despair. One allopathic physician lamented, "That which for the present, has a great though transient interest, seems to absorb the whole medical mind of the country; and, *docti indoctique scribimus*, of cholera! cholera!! cholera!!!—upon which no one sheds new light"

(AMA 1850, 107). Another offered this sober prediction: "Of cholera, it is probable, it [the specific cause] will never be known" (Seymour 1857, 188).

ALLOPATHY GETS AN ORGANIZATION

By midcentury, regulars faced an increasingly precarious professional position. While they had disassociated themselves from the excesses of rationalism, they suffered from an ascendant homeopathy and an inability to offer a coherent framing of cholera. Both of these problems could be traced, in part, to the embrace of radical empiricism. Eschewing theorizing and privileging local observation, regulars lacked standards by which to adjudicate competing claims. Intellectual fragmentation compromised their social standing. It also made them vulnerable fodder for homeopathic critique. Appealing to their own systematic empirical observations and the universal law of similars, homeopaths attacked regulars as unsystematic in their empiricism. Because regulars' growing suspicion of medical systems and rational theories precluded them from establishing general principles, "the old school is without a system of practice, or practice without a principle, and it is even a boast of its advocates, that there is no rule or law as a guiding principle in the application of remedies in disease" (Grabill 1857, 1). Or as another homeopath put it: "The physician remaining in the old school is bewildered with opposing theories and oppressed with an accumulation of heterogeneous and unarranged materials" (Sharp 1856, 98). These arguments gained traction. The public took a harsh view of allopathic accounts and treatments of cholera and sought out homeopathy as an alternative (Coulter 1973; Kaufman 1988). Particularly disconcerting for regulars was the support of homeopathy among the urban upper class (Coulter 1973). Emboldened by the repeal of licensing laws, homeopaths began to advocate for more inclusion in government institutions. One allopathic physician observed, "The Homeopaths are urging their claims to recognition on State and municipal boards whenever they can get an opportunity" (*Medical and Surgical Reporter* 1867, 16). Legislatures seemed open to homeopathic arguments, and by the end of the 1840s, many allopathic physicians were resigned to the reality of unwelcoming legislatures:

> The public, on the subject of medicine, intend well, but on everything connected with it they are lamentably ignorant. . . . Can we be surprised, then, that our legislators should be deluded into the endorsement of fantastic

> systems and modes of treatment, by the plausible assertions of cunning imposters, by partial and deceptive statements, the truth of which they have not the requisite knowledge to determine, and are compelled to take on trust? (Hutchinson 1867, 58)

Public "delusion" combined with the egalitarian ethos of the Jacksonian period ("*Equality* is the procrustean bed in which everything must be shaped" [Clark 1853, 272]) to make legislatures resistant to granting privileges to one sect over any others. "The history of the legislation touching the practice of physic and surgery" afforded a "melancholy illustration of the truth" of this reluctance (Hutchinson 1867, 56).

Regulars were losing the epistemic contest. Bewildered, they puzzled over their dwindling prestige: "It may, to say the least, be considered strange, that in almost every other department of knowledge besides that pertaining to the healing art, man seeks and follows with a degree of religious deference, the counsel and advice of those who are supposed to be the best informed and the most skillful in that particular branch which is his immediate concern; but in medicine it seems to be just the reverse" (Blatchford 1852, 70). To allopathic physicians, the widespread adoption of homeopathy represented a new low, a new nadir, in public common sense. "Men, in all ages, have been prone to trust to the absurd pretensions of empiricism in the treatment of disease, but never have intelligent men been so much disposed as at this day, to put confidence in the various sects of practitioners who do not profess to found their art on the science of which we are so proud, or even to be at all conversant with it" (Hun 1863, 6).

It was becoming clear to allopathic reformers that they needed a new strategy. Their intellectual arguments were not cutting it. Operating against a democratic headwind, they changed tack. Rather than attempt to win the intellectual/cultural debate with homeopathy, regulars adopted an organizational solution to their epistemological woes. They believed that the interests of allopathic medicine could be protected only through the formation of a national professional organization whose unity would filter down into local societies (Coulter 1973, 179). Through the establishment of a national medical society, allopathic physicians sought to reconstitute allopathy as a more coherent body, while simultaneously drawing firm boundaries between legitimate medical knowledge and quackery, to organizationally provide the very standards of adjudication absent in their radical empiricism.

In 1846, a small group of allopathic physicians met in New York to form such an organization. The idea of the national professional society was not in itself innovative; homeopaths founded the American Institute of Homeopathy (AIH) in 1844. But whereas the AIH was primarily a scientific society aimed at providing a forum for communication, the AMA was an exclusive society aimed at promoting allopathy at the expense of alternative medical sects. It was a conscious, unabashed tool for the promotion of regulars' professional interests. And the founders saw it as the beginning of a new era for allopathy. In the first address of the society, President Nathaniel Chapman (1848, 8) heralded the event as one that signaled that the profession had awoken "from the slumbers too long indulged," ready "to vindicate its rights, and redress its wrongs."

The driving issue for the founding of the AMA was the declining state of the profession (Rothstein 1992). Reformers identified three sources of this decline—an ignorant public, a weak educational system, and homeopathy—and decided to focus its energies on educating the public, reforming medical education, and combating quackery (Coulter 1973). As noted above, the programs of public education and educational reform took a backseat to the efforts to combat homeopathy (Coulter 1969). Because allopathic physicians understood appeals to the public as the sine qua non of quackery, they were reluctant to engage the public in any meaningful way, evidenced by the AMA's closed-door meetings and its lack of a public journal until 1883. Educational reforms were also stymied, often discussed but rarely acted upon, as reformers became mired in a stalemate with the proprietary medical schools, which resisted any attempts to increase educational standards (Ludmerer 1985; Marks and Beatty 1973; Rothstein 1992).

It was to the third task—combating quackery—to which the AMA turned its immediate attention (Fishbein 1947). By establishing the AMA, allopathic physicians sought to create organizational standards to adjudicate competing claims between legitimate medical knowledge and quackery, to draw a "strongly marked line of distinction between the educated and the uneducated, the liberal and the restrictive" (Brinsmade 1859, 22). The AMA's founders (Knight 1846, 750) argued, "In this state of things, the only resource which remains is, for medical men to establish and enforce among themselves such regulations as shall purify and elevate their own body, and thus more fully command the respect and confidence of their fellow men." The AMA sought to identify and eliminate irregular physicians through the "exercise of a moral power" (Hutchinson 1867, 59). If the state legislatures refused to officially recognize the difference between medicine and quackery, regulars would have to do it themselves.

Through the AMA, allopathic physicians sought to exclude homeopaths from the universe of legitimate knowers, turning inward to create their own system of regulation (Starr 1982). The main vehicle by which this was to be achieved was the codification and the enforcement of the 1847 Code of Ethics, specifically the no consultation clause (Rothstein 1992). The AMA restricted its membership to allopathic physicians who had allopathic medical training and rejected "unorthodox" teachings. Not only were homeopaths and other "irregular physicians" prohibited from joining allopathic medical societies, but allopathic physicians themselves were prohibited from seeking or giving consultations to alternative physicians:

> Although it is not in the power of physicians to prevent, or always to arrest, these delusions in their progress, yet it is incumbent on them, from their superior knowledge and better opportunities, as well as from their elevated vocation, steadily to refuse to extend to them the slightest countenance, still less support. (AMA 1851, 86)

The AMA was explicit about the rationale behind the no consultation clause; regulars needed to keep their "skirts clear of everything pertaining to irregular medical practices of whatever kind of description. We should not only avoid all complicity with them, but even the suspicion of a quasi recognition of them" (Hutchinson 1867, 62). Consultation was deemed as "giving 'aid and comfort to the enemy,' quackery, and as such is treason against the honorable profession of medicine" (Butler, Levis, and Butler 1861, 496). In presenting the code to the entire body, Dr. John Bell urged all members to "bear emphatic testimony against quackery in all its forms" (quoted in Fishbein 1947, 37). The intensity of this rhetoric underscored the commitment to defeating homeopathy, as did the severe punishment—expulsion—of those who indulged it. These exclusionary organizational practices represented a marked departure from the long tradition of interaction between the two groups and the recognition of professional equality in previous ethical codes of local medical societies. It raised the stakes of the epistemic contest for the average regular by officially drawing distinctions and prohibiting cooperation through various mechanisms of punishment and censure.

The Code of Ethics was strengthened in the years following the formation of the AMA. Before a local society could be admitted to the association, it was required to purge all homeopaths (Coulter 1973). In 1851, the no consultation clause was revised to explicitly grant the power to local societies to expel suspected homeopaths: "Each County meeting shall have the power to examine the case and immediately expel any member notoriously in the practice of Homeopathy, Hydropathy, any other form of quackery, without any formal trial, the same to be ratified by the succeeding Convention, any By-Law to the contrary notwithstanding" (quoted in Coulter 1973, 202). In 1854, the AMA, worried that local societies were not effectively policing their boundaries, set up a committee to inspect whether local members were still involved with irregulars. And in 1856, the AMA codified the illegitimacy of homeopathic knowledge by resolving that homeopathic works could no longer be discussed or reviewed in allopathic periodicals, effectively erasing any trace of homeopathic thought from allopathic discourse. The AMA showed great resolve in carrying out these threats, demonstrated best in 1884 when it responded to the New York State Medical Society's repudiation of the Code of Ethics by canceling the membership of most of its members (Burrow 1963, 20).

With these exclusionary practices, the AMA was able to police its members and to foster some cohesion by expelling those physicians who strayed too far afield. Such efforts were mirrored at the local level, with the instigation of the AMA. The example of the New York Academy of Medicine (NYAM) was indicative of this trend toward exclusion. Initially, the reaction of many New York regulars to homeopathy was one of curiosity and engagement, evidenced by the fact that the allopathic Medical Society of the County of New York awarded an honorary membership to Samuel Hahnemann in 1832 (Kaufman 1988). Rather than reforming this corrupted medical society, allopathic leaders in New York State, particularly New York City, decided to found NYAM. The old society was simply too catholic to meet the current challenges. In outlining their reasons for the establishment of NYAM, its founders listed the separation of regular practitioners from irregulars first (Van Ingen 1949). To accomplish this goal, the academy had to define what constituted an irregular physician. This was not a straightforward exercise in definition, for in New York, the boundary between homeopaths and regulars was fluid as they regularly consulted with each other. After much discussion NYAM adopted a broad rule that excluded "all homeopathic, hydropathic, chronothermal and botanic physicians, and also all mesmeric and clairvoyant pretenders to the healing art, and all others who at any time or on any pretext claimed peculiar merits for their mixed practices not founded on the best system of physiology and pathology, as taught in the best schools in Europe and America, and shall be deemed to exclude also all such persons as associate with them in consultation" (quoted in Van Ingen 1949, 13). The broad definition was adopted without a dissenting vote. Furthermore, to safeguard the society from homeopath infiltration, NYAM established a Committee of Admissions "to guard the portal of the Academy and to see that no irregular or unqualified practitioner gained entrance" (Van Ingen 1949, 8). As a result, the first fifty years of NYAM witnessed numerous accusations and purgings of homeopaths from its ranks. More energy was expended on these exclusionary actions than on any other activities.

The creation of an exclusionary association with rigid membership standards had two strategic benefits for the epistemic contest: it created an organizational and cultural space that denied the epistemic legitimacy of homeopathic claims and, in doing so, brought a measure of unity to allopathy. First, through the Code of Ethics, the AMA drew strict boundaries between homeopathy and allopathy, singling "out the sheep from the goats" (Sayre 1870, 55–56). In excluding homeopaths from consultations, allopathic physicians hoped to destroy public confidence in them, deprive them of clientele, and increase the gulf between homeopaths and allopathic physicians (Rothstein 1992, 171). In absence of firm epistemological standards upon which allopathic physicians could justify their rejection of homeopathic knowledge, regulars substituted membership restrictions, providing organizational standards for adjudication. Second, in defining the other, the AMA established an identity for allopathy despite its intellectual fragmentation. The defining feature of allopathy became opposition to homeopathic quackery. This new identity, rooted in an organization, mitigated some of the centrifugal force that accompanied allopathy's commitment to radical empiricism.[5]

Therefore, the AMA's epistemological accomplishment was not in facilitating a positive program of cohesive allopathic knowledge; it was in the formalization and institutionalization of the standing of homeopaths as quacks. While factions within allopathy had long been critical of homeopaths and homeopathic knowledge, boundaries between medical sects were not rigidly defined prior to the AMA. Homeopathic and allopathic physicians shared educational experiences. Many homeopathic converts maintained relationships with their regular colleagues after their defection, since, prior to the Code of Ethics, the stigma attached to such conversions was negligible. As competition between the sects intensified after the first two cholera epidemics, maintaining such openness and curiosity toward homeopathy became more difficult, but it nevertheless endured. The AMA changed this; it made it impossible to maintain one's status as a legitimate physician within the allopathic community while flirting (or giving the impression of flirting with) homeopaths or homeopathic ideas. To do so was to be painted with the debasing brush of quackery.

As homeopathy became institutionally deemed quackery, allopathy was not compelled to indulge it, even in order to refute it. It could be summarily dismissed as implausible, self-evidently false, and unworthy of attention. The surface ludicrousness of homeopathy did not mask some deep insight. The AMA had evaluated it and declared it foolishness. Prominent allopathic physician Worthington Hooker (1852, v), a self-appointed gadfly, articulated the AMA's position toward homeopathy:

> Absurd as Homeopathy appears on the face of it to the man of science or of plain common sense, the extent of its absurdity is revealed only by a thorough examination of its pretended facts and its plausible reasonings. . . . A wordy and finespun theory, built upon the loosest analogies, especially if accompanied, as is usual with all forms of delusion and quackery, with reports of wonderful cures, is sufficient to satisfy them, at least till some other system presents itself, with similar appliances for fascinating the ear of popular credulity.

As quackery, homeopathy failed to meet the most basic requirements of "common sense," succumbing to bald "delusion." Homeopathy was "a stupendous monument of human folly" and "a confused mass of rubbish" (Hooker 1849, 136). Another doctor wondered, "Is this not an incomprehensible science, indeed? Where is the mortal mind of capacity sufficient to grasp such a thought?" (Blatchford 1852, 88). Homeopathy, "a wordy finespun theory, built upon the loosest analogies," was the embodiment of outdated rationalism that allopathy had rightfully discarded.[6]

Allopaths claimed that the only explanation as to why anyone would adopt such manifest absurdities was "for the sake of money" (Hun 1863, 36). In the allopathic mind, homeopaths sought not to cure disease but to dupe the urban elite into paying exorbitant sums for their false therapeutics. It consisted of nothing more than "trickery, fraud, and chicanery, together with puffs of extraordinary cures, high pretensions to some peculiar power, some superior knowledge, and long catalogues of great and influential names of patients" (Hutchinson 1867,141) As such, homeopaths were lumped together with other confidence men of the time: "These ignorant tricking practitioners in medicine, constitute a race which may be classed with bogus jewelry pedlars , fraudulent lottery speculators, and all such like 'sharpers,' who take advantage of the simple-minded" (*Medical and Surgical Reporter* 1865, 214). Given its self-evident absurdity and the ignoble motives, allopathic physicians refused to engage with homeopathy in debates. They rejected any challenges from homeopaths to test their system and refused to take their claims regarding cholera seriously. Deemed quackery, homeopathy fell outside the realm of serious, legitimate knowledge. There was no need to indulge it, as it was just the latest in a long tradition of quackery, and when it fell out of favor, "something else equally absurd, (I cannot think of anything *more* so) will no doubt at once take its place" (Hutchinson 1867, 72–73).

It is important to point out that the allopathic rejection of homeopathy and its expulsion from the AMA were not determined by the content or nature of homeopathic evidence, but rather *by homeopaths' exclusion from the regular professional community*. Under these organizational standards, legitimate knowledge was to be judged according to *who* proclaimed it. If the doctor was a member of the AMA, his opinions could be considered. If not, they were dismissed outright as quackery. In other words, the distinction between quackery and legitimate knowledge was not based on particular epistemological or intellectual rationales. It was a distinction between legitimate and illegitimate knowers operationalized as membership in allopathic certified professional societies. Membership thus served as a solution to the problem of adjudication introduced by radical empiricism. The AMA established itself as a cohesive organizational community for allopathy, not by reconciling the mess of competing allopathic knowledge, but rather by defining a common other. Allopathic physicians were allopathic insofar as they were *not* homeopaths. The problems associated with allopathy's fragmented knowledge base were muted, relegated to internal communications, by the display of public unity of the profession through its professional associations. From this united front, allopathy waged its fight against homeopathy.

Because knowledge production is a collective endeavor (Longino 2002), the social organization of knowledge affects the knowledge produced (Shapin 2008; Shapin and Schaffer 1985). As an epistemic practice, the formation of the AMA, and the enforcement of its Code of Ethics, affected the content of allopathic knowledge on cholera in two ways. First, the restrictive membership practices limited the pool of perspectives regulars could bring to bear on cholera. As noted above, prior to the AMA, the boundary between allopathic and homeopathic physicians was blurry, especially since many homeopaths were allopathic converts who shared similar educational and professional experiences (Coulter 1973). Regulars often explored and even adopted homeopathic ideas. The AMA erected a rigid separation between the two sects. By expelling homeopaths from allopathic practice, barring consultations with them, and prohibiting discussion of homeopathy in allopathic journals, the AMA effectively eliminated *all* homeopathic knowledge from consideration. Homeopathic knowledge now fell outside of the "scheme of plausibility" (Shapin 1994, 22). Consequently, extended critiques of homeopathy like the famous one conducted by Oliver Wendell Holmes (1842)—critiques that required significant engagement with homeopathic ideas—were unfeasible after 1847. Allopathic physicians were prohibited from even acknowledging homeopathic ideas to refute them. The outright dismissal of homeopathy prohibited allopathy from making use of possible homeopathic insight into issues like cholera, or from strengthening its own arguments by debating homeopathic claims.

Second, the AMA's rigid demarcation encouraged the rejection of entire methodologies (e.g., provings and statistics) due to their association with homeopathy. For cholera, this was most evident in the allopathic rejection of statistical data, a rejection that represented a departure from the Paris School, which embraced quantitative data as a supplement to clinical observation (Ludmerer 1985; Warner 1998). As noted above, the rejection of statistics was driven as much by professional politics as it was by an epistemological rationale. In adopting a radical variation of empiricism, American reformers not only eschewed statistical data because, epistemologically, it abstracted knowledge from the local context of sensory observation and, in turn, ran the risk of devolving into the unwarranted speculative generalizations. They also rejected statistics because they were seen as homeopathic. This out-of-hand rejection of statistics precluded the use of statistical methodology to make sense of disease. Equating statistics with quackery, allopathic physicians rarely deployed them to study cholera, as evidenced by the dearth of statistical data in their discussions on the disease in their professional journals.

The AMA institutionalized the mistrust of homeopathy and fostered a general skepticism toward all knowledge that originated outside of the allopathic community. Despite the embrace of empiricism, the organizational restrictions on knowledge insisted upon by the AMA reinforced allopathy's long-held suspicion of appeals to the public and outside meddling. This was well illustrated by the fact that the AMA and NYAM held closed-door meetings, refusing to admit the press "since the public was not considered capable of understanding medical matters" (Van Ingen1949, v). By fomenting mistrust of outsiders, the AMA circumscribed the debates over cholera. Insofar as dialogic interaction is important for the growth of knowledge, this circumscription resulted in lost opportunities for allopathy to gain insight from outside sources, both methodologically in the rejection of certain ways of knowing and therapeutically in the dismissal of homeopathic cures for cholera, which might be more effective (or at least less harmful) in treating the disease (Coulter 1973; Kaufman 1988; Warner 1997). Once again, regulars, through the newly formed AMA, were left promoting an elitist epistemology shielded from potential outside revelations about cholera. Outsiders were not engaged; dialogues did not happen. Rather than partake in debates, allopathic physicians evaded them. They proffered an exclusionary space and dictated who was included as legitimate knowers. As such, any political benefit allopathy accrued from radical empiricism's ostensibly democratic character was undermined by its organizational policies.

The AMA's Coordinated Effort at Exclusion: Two Cases

Regulars tried to transform this organizational cohesion into a program aimed at preventing homeopaths from gaining government support or, failing that, preventing government recognition from being converted into tangible resources. No matter that doctors within the AMA held widely disparate views; this disparate group was unified primarily to combat a common other. Imparting a sense of cohesion and collective identity, the AMA attempted to use its organizational leverage to dictate to other institutions, especially state institutions, the universe of legitimate doctors as they defined it, so as to minimize homeopathic gains. The AMA punished allopathic physicians who worked alongside homeopaths and, in turn, regular practitioners, either willingly or unwillingly, avoided any association with homeopaths at all costs (Kaufman 1988). Any effort that included homeopaths was suspect in the eyes of the AMA, and in turn, dangerous for allopaths who participated. Conflicts therefore erupted around the government's various proposals to recognize and include homeopaths in state institutions. And while the government rarely recognized the exclusive claims of allopathy, the AMA was able to limit (or at least frustrate) all efforts of homeopaths to achieve equal standing within such institutions.

Two examples illustrate this strategy, one at the level of the state legislature, the other at the national level. The first example involved a proposal to include a homeopathic chair in the medical school of the University of Michigan (UM), a case that points to the ability of the AMA to assert itself through local societies to frustrate and delay homeopathic inclusion even in the face of legislative mandates. The 1849 epidemic created an opportunity for homeopaths in Michigan to advocate for inclusion in the medical school, which they seized. In 1855, in response to petitions from the state's homeopaths, the Michigan state legislature directed the university to establish, teach, and maintain the specific philosophies of both regular and homeopathic schools (Peckham 1994). The Michigan State Medical Society, with the support of the AMA, fought the provision. Citing the no consultation clause, it threatened the university's board of regents with decertification and the exclusion of any future graduates from membership in allopathic medical societies. Thus pressured, the board of regents refused to abide by the rule. This set off a decade-long debate over who ultimately controlled the University of Michigan, the state legislature (which supported homeopathic inclusion) or the board of regents (which supported the AMA position). In 1867, the legislature tied the establishment of a homeopathic chair to a new funding bill, funding sorely needed by the medical school. Once again, pressured by the AMA, the board of regents balked, offering a compromise solution by which a separate homeopathic school of medicine would be established in Detroit. Like many separate-but-equal compromises, this one stressed the former over the latter. Underfunded, the Detroit school split the homeopathic community, with some homeopaths supporting it and others viewing it as capitulation and continuing to advocate for inclusion at UM.

Finally, in 1875, the medical school, needing funding for a new hospital, agreed to establish two chairs of homeopathy, provided that homeopathic students received separate diplomas not signed by allopathic professors (Coulter 1969). The Michigan State Medical Society immediately adopted a resolution announcing a state of crisis between the society and the college, citing the potential retribution by the AMA on the college and its graduates (Haller 2005, 162). The AMA weighed in, stating that it would place the university under a ban, cease to recognize graduates from the college, and expel the college's professors from the AMA for violating the no consultation clause by teaching in the same college with homeopaths. After an intense three-year debate, the AMA decided against censure. In reviewing the consultation clause, the AMA ruled that it only applied to medical consultation, not membership on the same faculty. Still, the AMA turned its focus toward marginalizing the homeopathic professors within the college.

Although the AMA was only able to delay the inclusion of homeopaths in UM for two decades, it was more successful in preventing homeopaths from serving in the Union Army during the Civil War. With over three hundred thousand Union soldiers dying from injuries or disease, the Civil War "was a medical as well as human tragedy" (Ludmerer 1985, 9) that reverberated throughout American society (Faust 2009). This medical tragedy resulted in part from the woeful understaffing of the Union Army's medical department (Adams 1996). Yet despite the shortage, homeopaths willing to volunteer were excluded from service by the Army Medical Board, which was under the control of allopathy. In 1862, the Homeopathic Medical Society of Massachusetts petitioned Congress for the inclusion of homeopathy in every military hospital, citing cholera statistics from the 1832 and 1849 epidemics as part of their justification. Senator James Wilson Grimes of Iowa introduced Senate Bill 188, which would allow homeopathic physicians to serve in the army and would place some military hospitals under homeopathic control. For homeopaths the rationale was simple:

> Large numbers of Homeopathic physicians who are perfectly competent to care for the sick and wounded, practitioners who have been duly examined and recommended by Allopathic medical boards, and whose record in civil practice stands high and compares favorably, not to say enviably, with that of the Allopathic school, have been, and are, rejected on account of their medical *faith*. What seems stranger still, is the fact that thousands of those who are suffering *wish* the benefits of this practice. These sufferers are the very men who assisted in placing these public servants in office. They are fighting for a common cause, one in which they have as much interest as any other class. They fight, bleed, and die, not for monarchy, but for *democracy*. (Stow 1864, 257–258)

The bill passed, and was signed into law by Lincoln. The federal government gave the go-ahead to begin admitting homeopaths into the army.

Immediately, the AMA organized against the legislation. "The medical profession throughout the land, believing that such a measure would be detrimental to the life of the sick soldier, and that it would degrade and destroy the efficiency of the medical staff of the army, indignantly protest through their medical societies, medical journals and individually, against the enactment of such a law" (Medical Society of the State of New York 1867a, 65). Inclusion of homeopaths "would dissatisfy and dishearten the medical staff of the army, who understand the true character of homeopathy, and who have entered the service of their country with confidence that the government would strive to elevate the standard and promote the efficiency of the medical staff, results surely to be defeated by the appointment of Homeopaths" (NYAM 1862, 435). Controlling the Army Medical Board, allopathic physicians continued to refuse to admit homeopaths, even those who passed the army exam. Homeopaths admitted under false pretense, "would, if admitted, meet with a summary dismissal" (*American Medical Times* 1863, 59). The organized defiance succeeded, as the federal government was too preoccupied to enforce the Senate bill.[7] Homeopaths never served in the army during the Civil War.

In both of these cases, the AMA was able to undermine the inclusionary policies of government institutions albeit with different levels of success. The AMA, brandishing its no consultation clause, was able to rebuke or significantly delay government-mandated recognition of homeopathy, whether such government recognition was done in the name of equality, democracy, or even mere expediency. Drawing firm distinctions between the two sects, regulars appealed to the epistemic illegitimacy of homeopathy, its quackery, to justify their refusal to cooperate with such mandates. The varying degrees of success of these efforts were determined by the degree to which the AMA could exert its control over public institutions. In the case of the University of Michigan, this control extended only so far, as the legislature took an active role in the controversy to promote the equal treatment of homeopaths as codified in the state's statutes. In the Union Army, allopathic control of the Army Medical Board, coupled with the lack of government oversight of the board, allowed the AMA to successfully prevent homeopaths from achieving equal status. By pressuring its members, the AMA ensured that the formal, legal recognition of homeopathic claims of equality did not translate into tangible resources.

CONCLUSION: ORGANIZING FOR EPISTEMOLOGY

The founding of the AMA, and the particular policies it adopted, were shaped by the epistemic contest. Needing to adjudicate between good allopathic knowledge and impoverished homeopathic quackery, and lacking the epistemological standards to do so, regulars adopted an organizational solution. This chapter illustrates that epistemic contests are waged by a variety of means, involving both cultural and organizational strategies. Indeed, precisely because epistemic contests lack basic *cultural* agreement on the ideals of knowledge, organizational strategies assume great importance. To myopically focus on cultural strategies as the means to resolve such intellectual disputes obscures the rich repertoires actors deploy in advocating for their visions of knowledge. And unduly restricting one's analysis to *either* organizational or cultural factors is to buy into a false choice driven more by theoretical biases than historical and empirical imperatives. Epistemic contests can be (and often are) adjudicated, not through specific debates over the merits of some form of knowledge over others, but through organizations, which set the parameters of the debate and determine who can and can't engage in it, or even whether the debate can occur at all. Advocates of epistemological positions negotiate this organizational terrain, attempting to harness organizations in the service of their epistemological ends.

Hampered by a lack of well-defined epistemological standards and ineffectual intellectual arguments, allopathic physicians adopted an organizational strategy to solve the problem of adjudication and carry out a program of exclusion. How successful was this strategy? Beyond the specific outcomes of the University of Michigan and the Union Army, it had mixed effects on the professional project of allopathy. On one hand, the AMA imposed some intellectual order on a situation lacking firm epistemological standards. Through membership restrictions and the no consultation clause, the AMA introduced organizational criteria by which to adjudicate competing knowledge claims, something not possible within the confines of radical empiricism. Defining homeopathy as quackery, it successfully prohibited homeopathic thought from gaining a foothold within allopathic circles. It established a firm boundary between allopathy and homeopathy, registering important distinctions, an invaluable contribution in the messy context of nineteenth-century medical practice. And while the exclusionary membership policies did not solve internal fragmentation, they did bring a measure of organizational cohesion to allopathy. In essence, it enabled the development of a "thought collective" (Fleck 1979) with only a modicum of shared thought, allowing allopathic physicians to wage the epistemic contest with homeopathy in spite of its own intellectual fragmentation. Allopathic physicians avoided cultural/intellectual debates with homeopaths by deeming them as falling outside of the realm of legitimate knowers, unworthy of engagement. Instead, they shifted the epistemic contest to an organizational terrain where they had more leverage.

On the other hand, the early AMA's exclusionary practices circumscribed allopathic approaches to knowing in a way that ensured a continued degree of internal intellectual fragmentation. By artificially restricting the parameters of the debate, the AMA encouraged the summary dismissal of methodological and intellectual insights associated with nonallopathic sources (e.g., statistical methods) that may have helped resolve some of these internal debates. And as much as the AMA tried to contain these debates within the confines of the organization, they led to incoherent public appeals on pressing issues like cholera, especially in opposition to the neat statistical rhetoric of homeopaths.

Perhaps more important, the strategy of exclusion was discordant with the democratic spirit of the time, as it sought to restrict, rather than promote, dialogue and openness. In adopting this strategy, allopaths forfeited any claim to a democracy that was part of the original allure of radical empiricism. Homeopaths picked up on this, adopting antimonopolistic rhetoric reminiscent of Thomsonism. They argued that the AMA, in its efforts to exclude homeopaths, was trying to ensure that the practice of medicine was determined "by *creed*, not *fitness*" (Homeopathic Medical Society of the State of New York 1874, 27). According to homeopaths, allopathic physicians were content to "call names, make faces and throw stones at opponents whom they dare not face in a fair field" (Cornell 1868, 4). Avowing that "the medical profession is not an aristocracy created for the benefit of a caste" (Bowers 1871, 112), homeopaths portrayed themselves as champions of openness and democracy to great effect. Taking New York as one example, we see that the legislature continuously legislated in homeopaths' favor throughout the nineteenth century (see table 2.1). The AMA, and its local satellites like the New York Academy of Medicine, repeatedly failed to make a convincing case for the exclusion of homeopaths to the legislature. This failure resulted in part from the lack of intellectual coherence that plagued radical empiricism.

Table 2.1. Continual allopathic defeat in the New York State Legislature

Year	Legislative outcome
1844	The 1830 licensing law is repealed. Homeopathy and other alternative medical sects are allowed to practice medicine.
1857	Legislature charters state and county homeopathic societies to issue voluntary licenses for practice, granting equal legal status to homeopathic professional societies.
1866	Homeopath physicians are granted membership on the newly formed Municipal Board of Health of New York City.
1872	Legislature approves a noncompulsory exam program proposed by homeopaths, rejecting an allopathic proposal.
1880	Legislature broadly defines a physician as a graduate from a legally incorporated medical college (allopathic, homeopathic, or eclectic) who has registered his diploma in the clerk's office in his local county.
1890	A new licensing law is passed, which establishes three separate licensing boards for each school of medicine (allopathy, homeopathy, and eclecticism), rather than allopathy's plan for a single exam board with an allopathic majority. Final vote was 109 to 4.

But it also stemmed from the antidemocratic critiques to which allopathy was especially vulnerable. The AMA's campaign of exclusion did little to assuage allopaths' antidemocratic reputation.

In the end, the early actions of the AMA failed to radically change the general trajectory of the epistemic contest over medicine. But this is not to say that it was a complete failure or that its early activities should be dismissed as ineffectual and thus irrelevant. While the formation of the AMA did not translate into an immediate professional victory for allopathy, it was an important event in the process of professional consolidation and epistemic closure, as it established the organizational terrain of the epistemic contest for the rest of the century, shaping the terms of the epistemological debate and the subsequent knowledge produced in the context of this epistemic contest. From 1848 on, the sorting out of cholera, and the painful struggle that followed from it, would have to go through the AMA.

3

THE INTELLECTUAL POLITICS OF FILTH

IN 1866, THE UNITED STATES BRACED ITSELF FOR ANOTHER cholera epidemic. The disease had threatened invasion for a number of years—an ominous prospect for a country wrought by civil war. The war's dramatic pageant of death claimed over 620,000 soldiers plus an undetermined number of civilians (Faust 2009). The country, mired in turmoil, was also deep in mourning. The sheer scope of the war's mortality—it claimed nearly 2 percent of the U.S. population—altered the country's relationship to death, creating what Frederick Law Olmstead called "a republic of suffering" (quoted in Faust 2009, xiii). Such a republic was in no position to deal with another cholera outbreak. Fortunately, cholera waited almost exactly a year to the day from Lee's surrender at Appomattox before reappearing on U.S. shores. The battered country did not have to contend with cholera during combat. A small solace, but a solace nonetheless.

In 1866, cholera still induced fear. Thirty years after the first epidemic, doctors and patients alike vividly recalled the bluish pallor, the aching muscle cramps, and the speed at which the disease claimed its victims. However, unlike the previous two U.S. epidemics, this fear was tempered by familiarity. Living memory blunted the more exaggerated terrors. Cholera had lost some of its foreign mystique, and the previous epidemics, though serious, never reached the dire levels predicted by the panic-stricken press.

Familiarity may have bred some comfort, but much of the cautious optimism had another source. Due to promising developments in sanitary science prior to 1866, there was hope that this time officials might be able to prevent an epidemic, or at least contain its effects. A diverse group of public health reformers argued something unthinkable during 1832 and 1849—that cholera was, in fact, an easily manageable disease. They contended that if localities instituted a few key preventive sanitary measures, they could sit by and watch as the disease harmlessly passed by their communities. The premise underlying the optimism of the sanitary movement was that disease was filth. The *Sanitarian* (1873, 222–223), a major organ for the new sanitary movement, proclaimed unequivocally, "No truth is better established than that *dirt* and *impurity* are potent instrumentalities for the propagation of cholera. . . . Want of pure air, want of pure water, want of drains, want of sewers, and in cities want of salubrious exercise grounds, combine to produce dampness, stagnation, uncleanliness, misery, disease and death." The lesson to be drawn from the localization of disease was simple, yet revolutionary in its effects: clean up filth, eliminate cholera.

This is not to suggest that the arrival of cholera was taken lightly. The optimism was *very* cautious, uncertainty still abundant. Indeed, in 1866, Dr. W. C. Roberts of the New York Academy of Medicine warned against excessive confidence, reminding his audience that cholera was still "the one great epidemic. None of the diseases which have at different times ravaged the earth have equaled it in extent, rapidity, and simultaneousness of progress, nor [*sic*] in mortality" (NYAM 1871, 37). President of the Albany Medical Society James M'Naughton (1852, 126) endorsed this caution: "Familiarity [with cholera] has not divested this formidable scourge of its interest, or of its terrors. It is deservedly regarded as one of the most destructive diseases, by which the destroying angel has ever executed its commission." Hoping for the best, but expecting the worst, the public still understood cholera to be a grave threat.

As in past cholera epidemics, attention turned to New York City, cholera's favorite port of entry. In January 1866, the *Nation* (1866, 40) reported that it "was generally expected by men of science, as well as by the public at large, not only that we shall have a visit from the cholera during the coming spring, but that it will first show itself in New York, and commit here its worst ravages, and from this spread itself over the country, along every line of railroad." Aware of cholera's typical course, New York City officials took proactive measures. On February 26, 1866, after a decade of stunted reform efforts, reformers had secured the passage of the Metropolitan Health Bill, establishing the country's first permanent and politically independent municipal board of health. It was hoped that the newly formed board, insulated from city politics and the pernicious influence of patronage, would finally clean up the city and prevent the spread of cholera. Without delay, the board went to work readying the city. Between March 2 and March 7, the board issued a dizzying amount of decrees in a coordinated effort to clean up the city (Duffy 1974, 3). Police were instructed to submit weekly reports of all instances of filth on streets, wharves, and piers; physicians were directed to report all cases of contagious infections to the board; and warnings were sent to all owners and landlords that all establishments risked demolition if not cleaned. Taking on entrenched political interests and city graft, the board also ordered a review of all of the city's sanitary contracts and threatened to void the contracts of those who were not honoring them (Duffy 1974).

Still, anxieties would not be mollified until the board proved its mettle during an actual epidemic. Cholera would be the board's first test. Reeling from overcrowding, draft riots, and general urban unrest, the situation in the city remained precarious, and New Yorkers waited with bated breath to see if the new Metropolitan Board of Health could overcome the impending epidemic. The delay of the state legislature in passing the Metropolitan Health Bill compounded the general angst as many questioned the board's preparedness; while the board struggled "with vigor" to quickly carry out its reforms, "it was late in the season for much of their work, and Summer will be upon us probably before the most important part has been completed" (*New York Times* April 9, 1866, 4).

As cholera neared, the board intensified its efforts. It hired thirty temporary assistants to carry out a comprehensive sanitary audit of the city, ordering them to locate and then remove the most egregious offenses. A citywide cleaning effort got under way, a welcomed departure from the previous sanitary regime, politically connected and hygienically neglectful. By April 9, the *New York Times* was predicting that even if the board could not prevent cholera, it should at least curb its excesses (Duffy 1974). The paper also encouraged the board to maintain its nerve in the face of the "powerful and selfish interests which are trying to perpetuate the nuisances and causes of disease in our City" (*New York Times*, April 9, 1866, 4).

On April 13, cholera arrived. The board immediately lobbied the governor's office to proclaim a state of emergency and extend its already substantial power. Granted this unprecedented authority by Governor Reuben Fenton's poetically named "Proclamation of Peril," the board, with the full assistance of the Metropolitan Police, carried out an ambitious emergency program. It eliminated so-called cholera nests, isolated affected individuals and establishments, strengthened and maintained a quarantine system, and cleaned up the city's sewers, tenements, and other nuisances. Medical care was organized around house-to-house visits. Once indentified, the sick were transported to one of the board's six dispensaries or cholera hospitals established specifically for the epidemic (Duffy 1974). Never had the city coordinated such an extensive medical campaign.

Cholera arrived and departed with little more than a whimper as it was largely contained to the port. Only six hundred New Yorkers died out of a population of over eight hundred thousand. The board had passed its first test with aplomb. This is not to suggest it was easy or without controversy. The board faced stiff local resistance in every neighborhood in which it sought to establish a cholera hospital. Still, despite these moments of public outrage *during* the epidemic, assessments of the board were almost universally complimentary *after* it. The final verdict came a year later, when the *New York Times* (March 31, 1867, 3) concluded, "It ought to be permanently remembered to the credit of the Board, that having to deal thus early with the epidemic, they succeeded in checking its progress." Regardless of whether it was actually the sanitary reforms that led to the decreased mortality or simply a milder form of the disease, the Metropolitan Board of Health of New York City received the credit. It had tamed cholera, and, in doing so, offered other cities and communities a blueprint, which many began to implement shortly thereafter. With their proliferation throughout the country, boards of health became the new front in the epistemic contest over medicine, as multiple actors salivated over their extensive resources and power.

FRAMING EPISTEMIC AUTHORITY

The establishment of the Metropolitan Board of Health was a watershed moment, both in the U.S. experience with cholera and in the history of American medicine. Not only did the board usher in a period of interventionist sanitary reforms that would eventually conquer cholera; it also embodied a shift to a new secular conceptualization of disease, marking the date when cholera became a "social" rather than "spiritual" problem (Rosenberg 1987b). Cholera as a scourge from God could not be combated by mere human measures; cholera as filth could. The arsenal against the disease now included more than just prayer.

Yet the understanding of cholera as a *social* problem did little to mute the *medical* debates surrounding the disease. An important event, the establishment of the boards of health did little to resolve the epistemic contest. In fact, it stoked the epistemic contest by introducing a new type of organization with resources and power to fight over. As the boards of health became *the* main organization body that responded to cholera, they became an alluring prize to be won for allopathic physicians desiring to promote their professional goals. And the public esteem granted to the boards in the wake of the 1866 epidemic sweetened their allure.

The boards also introduced some complicating factors into the epistemic contest for allopaths. Prior to the boards' establishment, regulars focused mainly on ensuring homeopaths' marginalization. The boards, and the sanitary reform movement that underwrote their establishment, brought new players into the epistemic contest, all of whom sought the recognition to speak authoritatively on cholera and disease in general. A motley group of reformers, the "sanitarians" included influential community members, civil engineers, plumbers, progressive politicians, and physicians from all sects (Rosenkrantz 1974).[1] The new actors each staked a claim to defining cholera in order to achieve different ends. Politicians sought patronage and potential kickbacks that accompanied a new institution with significant resources. Political reformers sought to use the boards as a model for their program to introduce rationality and integrity into municipal government. Homeopaths and other professionals (e.g., plumbers) sought inclusion on the boards and recognition for their essential contributions to the elimination of filth. And regulars tried to harness the boards to achieve professional and epistemic authority by taking credit for their sanitary successes. With all these competing actors and interests converging in the boards of health, the boards became not only the primary organizations to deal with cholera; they also became important sites for the epistemic contest over medical knowledge, ownership over the understanding and definition of disease, and the means by which society was to intervene. The boards became contentious arenas in which diverse actors asserted their status as privileged knowers and claimed epistemic authority.

This chapter investigates the consequences that the establishment of local boards of health, specifically New York City's Metropolitan Board of Health, had for the epistemic contest. During the 1860s, opinions about cholera (allopathic, homeopathic, and lay) converged around the notion of cholera as filth, to be eradicated through municipal reform. A coalition of lay elites and sanitary-minded physicians led the calls for sanitary reform. In the process, they framed epistemic authority in terms of its disinterested, apolitical nature, juxtaposing it to corrupt city politics. Expertise was seen as emanating from a particular ethos toward knowledge. As such, the boards promoted a type of "intellectual ecumenism" when it came to knowledge on cholera, one that encompassed not only medical knowledge but also a whole host of other forms of relevant knowledge.

As discussed in the introduction, epistemic authority can be justified along a number of interrelated but distinct dimensions. *How* actors frame their arguments for epistemic authority matters, as it delineates the types of arguments that can be legitimately mustered in defense of knowledge claims and dictates the organizational responses to these claims. Among the many ways in which epistemic authority can conceivably be justified, three are most common. First, such authority can be grounded in the *content* of one's knowledge. Recognition as a privileged knower is seen as deriving from the possession of a specialized body of knowledge. Claims made by clergy, which are based upon their understanding of sacred texts and a special spiritual insight, typically assume this form; clergy *possess* spiritual insight unavailable to the laity. Second, epistemic authority can be claimed on *methodological* grounds. Science is often legitimated this way. Scientists cannot be granted authority on the basis of the factual knowledge they possess, for the body of scientific knowledge is forever evolving. Rather, they claim epistemic authority based on their ability to achieve knowledge through the scientific method. Methodological appeals can be made in the name of an abstract ideal, like the scientific method, technical acumen (e.g., IT personnel), an ability to gather and process information (e.g., journalists), or the capacity to translate knowledge into practical applications (e.g., engineers). Finally, epistemic authority can be justified along the lines of an *ethos* or orientation toward knowledge. In these cases, it is not in the mechanics of knowledge production or the final knowledge produced that epistemic authority is justified, but in the stance one assumes toward the production of knowledge. An ethos-based approach to epistemic authority tends to place great emphasis on the character or position of the knower.

Sanitarians justified their epistemic authority in this final fashion, grounding it in their particular orientation vis-à-vis knowledge. Specifically, they claimed a special accuracy for their knowledge on account of its disinterested, apolitical character. The first section of this chapter describes this framing strategy by sanitarians in New York City, showing how it enabled them to convince the New York State legislature to establish a permanent, politically independent board of health. The decision to frame their epistemic authority along the lines of an apolitical ethos emerged in part from their opposition to city politicians. Because sanitarians approached sanitary knowledge without any political stakes in the findings, they could produce sober assessments of the city and yield rational, effective interventions to combat cholera. They were disinterested and, thus, their knowledge was more accurate than that of politicians, who had incentives (i.e., political patronage) to produce faulty knowledge that masked the extent of sanitary problems. Both sanitarians and their opponents in the City Inspector's Office used similar methods to gain knowledge of disease, but sanitarians claimed that the knowledge they achieved through such methods was more trustworthy, not because they possessed superior technical know-how, but because they were untainted by political calculations.

This particular framing not only emerged as a reaction to corrupt politicians but also flowed from the nature of sanitary knowledge itself. The diverse nature of the content of sanitary science prevented sanitarians from making claims based on the *content* of any one specific body of knowledge. New techniques like dot maps, normal mortality ratios, and sanitary surveys linked cholera to place and, in turn, recognized the relevancy of different types of knowledge related to place. With disease framed broadly as filth, possible relevant knowledge was wide-ranging; sanitarians pooled insight from a variety of sources. Medical knowledge was important, but so was engineering, legal reasoning, architecture, and even plumbing. The broad manner in which disease was defined necessitated a sanitary science that was practical in orientation and ecumenical in nature. Appeals to epistemic authority on specific technical and methodological grounds were not possible when so many different forms of expertise were deemed relevant. Rather, the new boards of health were conceived as a reform movement that targeted political corruption through an appeal to a disinterested ethos toward sanitary knowledge.

In terms of the epistemic contest, the ecumenical nature of the boards was detrimental to regulars' professional goals as it frustrated their attempts to gain control over public health. The second section of this chapter discusses the internal struggles over control of the Metropolitan Board of Health first between regulars and other sanitarians, specifically plumbers, and then between regulars and homeopaths. Allopathic physicians could not exclude nonmedical experts because the broad framing of disease as filth required the input of a number of actors, everyone from civil engineers to plumbers. They also were unable to exclude homeopaths from the board, for the expansive understanding of relevant knowledge translated into a broad recognition of relevant *medical* knowledge. Compounding matters, every attempt on the part of regulars to control the boards was effectively discredited by opponents as crassly political and in direct opposition to the stated ethos of the boards. Regulars' attempts to justify their control of the boards were therefore hampered by the manner in which the epistemic authority of the board was framed. As a result, they failed to turn the boards to their professional advantage.

As such, allopaths developed an ambivalence toward public health generally. On the one hand, the popularity of the boards made them a potential resource for bolstering the prestige of allopathy. Insofar as the sanitary measures of the boards were seen as successful, regulars benefited from an association with them. On the other hand, the intellectual ecumenism of the boards, and sanitary science more generally, made it very difficult for regulars to control the agenda of public health. Allopathic physicians found it difficult to assert the superiority of their medical knowledge over other forms of relevant knowledge (e.g., engineering, plumbing, etc.) and, even, over medical knowledge itself. In the end, the boards became part of the problem, not the solution, for regulars' professional aspirations. Once seen as prizes, they became yet another government agency to be viewed with suspicion.

LOCALIZING CHOLERA AS FILTH

By 1866, the confused debates as to the nature of cholera persisted, but most groups vying for control over the definition of cholera had reached agreement over a single fact—cholera was somehow related to filth. Historians have tended to treat this as an indicator of an emerging consensus for the miasmatic theory of disease (Barnes 1995; Duffy 1990; Leavitt 1992; Mitman and Numbers, 2003; Richmond 1947, 1954; Rosen 1993; Susser and Susser 1996). But the convergence of opinion was in fact more complex. The etiology of the disease and its relationship to filthy conditions remained a point of contention. Did filth create cholera? Did it just facilitate its spread by providing a fertile environment for growth? Or did filth simply undermine the health of the inhabitants living in it, making them more susceptible to the disease? Despite these persistent questions about the mechanisms behind the localization of disease, there was a degree of interpretive flexibility inherent in it. Some medical thinkers undoubtedly equated cholera with filth, but most adopted the more modest interpretation that there was some sort of demonstrable relationship between the two. Cholera need not be caused by filth in order for it to be related. Filthy locales were dangerous (Humphreys 2002), but how and why need not be specified. Minimally, all that the association of cholera with filth demanded was an acknowledgment of *some sort* of connection between place and disease for a wide array of actors to close ranks around commonsense sanitary interventions And it was flexible enough to allow for the commitment to it from a variety of epistemological perspectives, but definite enough to provide a common ground for disparate actors to coalesce into a public health movement. The theoretical stakes were minimal, but the practical implications great.

Beginning in the 1850s, the association of cholera and filth was given a concrete form by three new techniques—dot maps, sanitary surveys, and the statistical artifact of "normal mortality." In many ways, this association required no mean feat of investigation. The copresence of squalor, filth, and disease was apparent to all familiar with certain urban streets. The *New York Times* (June 25, 1856, 3) noted that just walking down the street was enough for someone to "consent that the *sense of smelling* itself is a nuisance." What the new techniques offered sanitary reformers was the elaboration and systemization of such anecdotal, sensory observations. Less important as sources of new insight, they served to illustrate existing ideas, providing visual representations that could be rhetorically deployed in the advocacy of sanitary reform. And sanitarians used them to great effect.

Mapping Disease—Even though the perceived link between environment and disease was not new to the nineteenth century,[2] doctors and sanitarians began to illustrate this link in the mid-1800s using maps. While most early maps of cholera were European (especially English) in origin, by the 1850s, Americans were involved in mapping disease (Barrett 1996). The growing prominence of medical mapping received great impetus from epidemic diseases that emerged during this period (Jarcho 1970, 138), especially cholera (Vinten-Johansen et al. 2003, 322). This interest in mapping epidemic diseases coincided with technical advances in printing, specifically the shift from copperplate to lithography, which allowed for more elaborate maps and more efficient dissemination of these maps (Koch 2005, 41). In this "golden age of medical cartography" (Gilbert 1958, 173), sanitary reformers adapted maps not only to make sense of cholera but also to illustrate particular arguments as to its nature.

During this period, two types of cholera maps were produced. The earliest maps were progress maps that traced the spread of cholera across large areas (Vinten-Johansen et al. 2003, 323), visually representing the movement of cholera *through* space. They emphasized the mobility of disease. As such, progress maps promoted a view of cholera more favorable to contagion theories. Temporality was an important dimension, as dates of outbreaks in specific areas were often noted along the routes that cholera took. However, because they covered wide geographical spaces, progress maps were limited by certain informational barriers; it was difficult to accumulate accurate information from disparate geographic locations. Since the United States lacked an adequate infrastructure for the collection and aggregation of vital statistics, the data behind progress maps lacked the requisite reliability, and as such, the maps had limited impact.

Of much greater importance were the dot maps, or spot maps, of cholera. In the mid-1800s, cartographers began to focus on the local circumstances of areas with the highest incidence of cholera. The most famous of these nineteenth-century dot maps was John Snow's map of the 1854 cholera outbreak in London.[3] Unlike progress maps, the intent of dot maps was to show the clustering of disease *in* space. Cartographers mapped out a geographical area, and then using "dots" or some other marking device, noted the incidence of disease. Clusters of disease reflected unhealthy local conditions.[4] As techniques developed, they used more sophisticated techniques for "spotting" their maps, such as shading techniques to visually relate the incidence of cases to population density. Interpretively, dot maps suggested a robust correlation between place and disease.

Like all maps, the dot maps were not mere presentations of facts. They marshaled selected propositions into arguments about the nature of the disease (Koch 2005). In other words, the cholera dot maps were arguments in visual form, and the argument they sought to convey was that there was a relationship between disease and some local factor (e.g., nuisances, cesspools, inadequate sanitation). For example, Snow's map linked disease to contaminated water sources, most famously the Broad Street pump (Johnson 2006). Indeed, it was this ability to visually convey information that was most alluring about the maps. Complex, poorly understood disease properties were reduced to dots on a map—dots when aggregated conveyed a pattern not readily "seen" otherwise.

For sanitarians maps performed the crucial tasks of simplification, reduction, and translation. And they interacted with the viewers in a seemingly transparent way. Knowledge claims were made visible and legible to the public. Like statistics for homeopaths, maps made arguments through demonstrations of knowledge, albeit in visual form. In this respect, they sug-

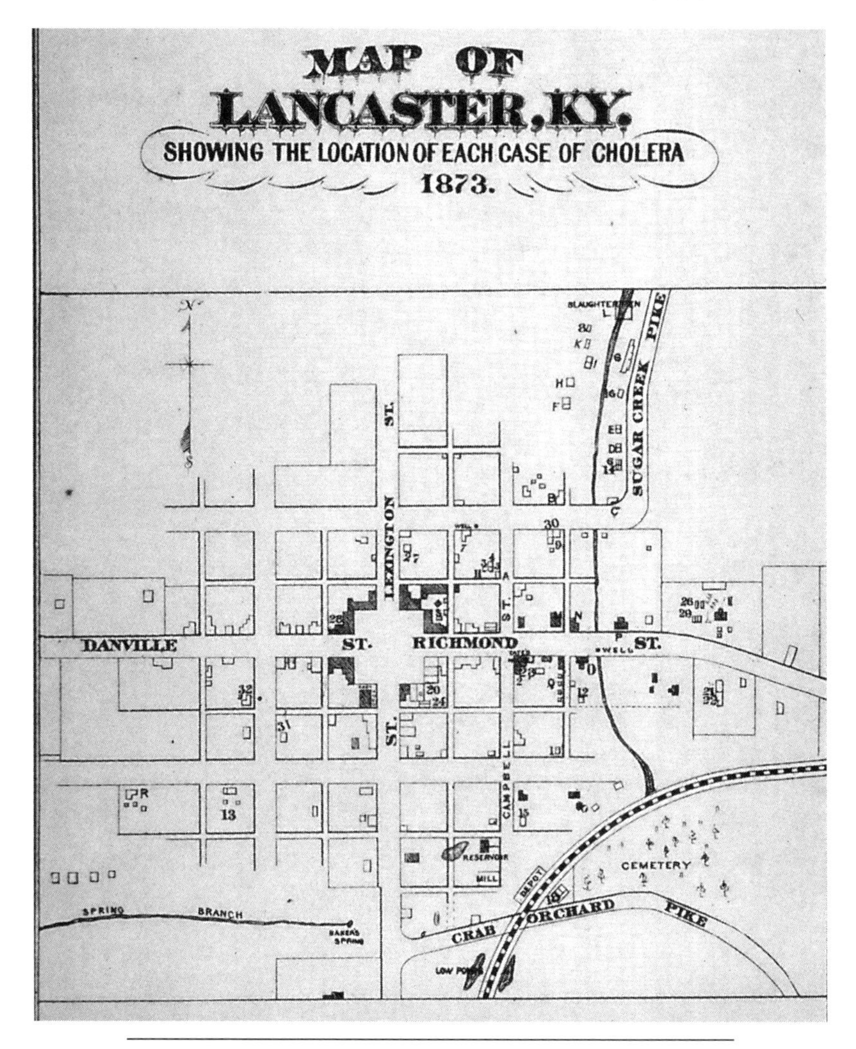

This dot map indicates every cholera case (represented as black boxes). *Map of Lancaster, Kentucky, Showing the Location of Each Cholera Case in 1873*. Courtesy of the National Library of Medicine.

gested a more democratic form of knowing, even if this democratic commitment existed more in appearance than substantively. Indeed, as arguments in visual form, the maps excluded and masked as much as they included and revealed. Still, they made knowledge *seem* transparent and open to public evaluation. It is the appearance of transparency, and the unconscious way in which maps insinuate arguments (Boggs 1947), that make them effective

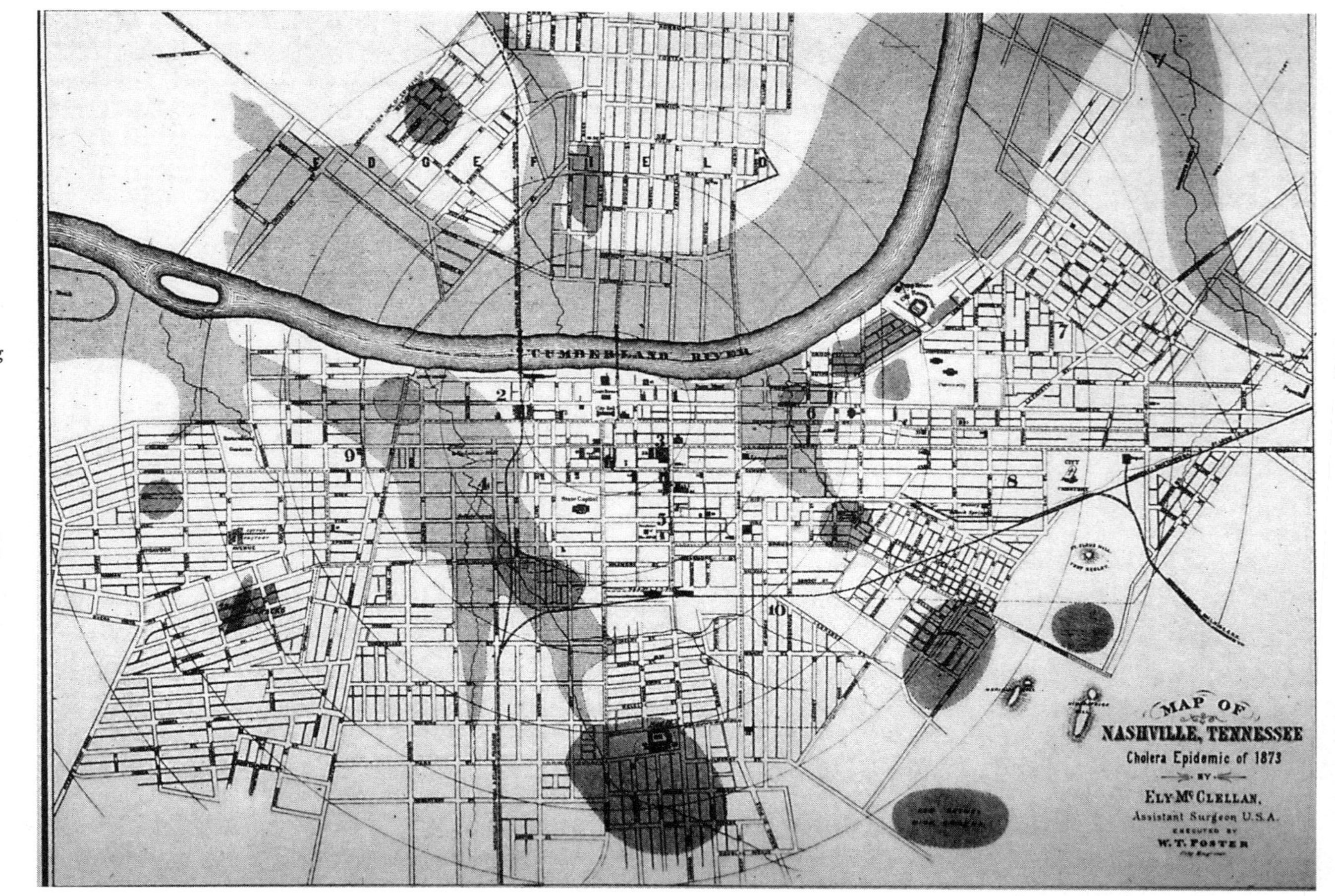

This spot map illustrates the more advanced mapping technique of shading, showing the varying incidences of cholera outbreaks through shading (the darker the shading, the higher the incidence) in Nashville, Tennessee, from Ely McClellan, 1873. *Map of the Cholera Epidemic in Nashville, Tenn., in 1873.* Courtesy of the National Library of Medicine.

rhetorically. Dot maps took complex arguments about the nature of cholera and transformed them into simple visual representations, rendering them perceptible to all.

The message dot maps conveyed was clear: when it came to disease, place mattered. While neither inherently contagionist nor noncontagionist, dot maps fostered a certain localism by offering a static picture of cholera (Stevenson 1965). By encouraging an "inherently ecological" way of thinking (Koch 2005, 2), they focused viewers' attention to the environmental context of disease. They established a spatial relationship of disease and some other local factor (a source of putrefaction, a ship or a pier), rooting a disease in place by relating it to fixed characteristics of the environment. Consequently, while the maps did not exclude any notion of contagion, these notions would have had to be actively read onto the visual representation. They were not inherently noncontagionist, but they made contagion more difficult to see.

Normal Mortality—A great obstacle for medical cartographers was gathering accurate knowledge to collate and fashion into useful maps (Osborne 2000). Maps were joined with and backed by vital statistics. Indeed, historically the emergence of medical mapping coincided with the development of medical statistics, as both shared a common intellectual base as part of the slow progression of medical science (Koch 2005, 8). But more than just gathering raw data, cartographers had to construct statistical techniques and measures to impose order on such data.

Sanitarians drew on statistics to support the visual arguments of the dot maps. It was not enough to show that diseases clustered in some areas more than others; cartographers also needed to show that this clustering was somehow exceptional. After all, people naturally had to die somewhere. To indicate the presence of preventable deaths, sanitarians developed a statistical artifact—the "uniform law of mortality" (Shattuck 1850, 95) or the "normal death-rate" (Smith 1911, 119). This measure was used to support the argument that certain sanitary locales were unnatural as they produced outcomes that violated the normal incidence of mortality. In essence, American sanitarians cobbled together the limited available vital statistics and calculated the mortality rate due to "inevitable causes," like old age, accidents, and endemic disease. This was deemed the normal death rate. Deaths from epidemic diseases fell outside the normal death rate, and, in turn, were by definition preventable. Through neat, obfuscating data management, sanitarians reframed deaths from cholera as aberrations to be prevented by the proper application of sanitary reforms, rather than part of the normal course of nineteenth-century living.

With the concept of normal mortality underlying their construction, dot maps became visual representations of nonnormal mortality that was linked to filthy local conditions. Rhetorically, this statistical ratio possessed all of the benefits—reduction, simplicity, and mobility—that the homeopathic statistics had. The measure effectively transformed the central message of the sanitary movement—sanitation could prevent unnecessary death—into a simple, seemingly transparent numerical ratio.

Sanitary Surveys—The normal death rate supplied an abstract, rough numerical indicator of the pernicious effect of poor sanitation on health. It quantified disease abnormality but did so via abstraction from the reality on the ground. Behind this abstraction, however, was the mundane activity of data collection. To create dot maps and to calculate normal death rates, reformers had to acquire hard data on local conditions that would explain the variation in mortality rates and the patterns observable on dot maps, to pinpoint the conditions of filth that lead to nonnormal death. Sanitarians conducted sanitary surveys that not only provided important information but also became their most effective rhetorical tool.

Sanitary surveys required good old-fashion "shoe-leather" research to gather local intelligence. Indeed, America's most important contribution to medical cartography was the idea that the local conditions of disease could not be gleaned through correspondence or from decontextualized statistical tables. Researchers had to go to the place to understand the relationship between disease and place (Barrett 1996). To produce their sanitary surveys, sanitarians divided a geographic locale into discrete sections and then sent surveyors to note the conditions on the ground and take inventory of a plethora of environmental factors. The comprehensiveness of the sanitary surveys reflected sanitarians' desire to reconstruct the total environment of a specific locale, because the total environment might be relevant to the prevalence of disease. These diverse conditions were then correlated with "prevailing sickness and disease" in all the specific locales that when amassed provided a comprehensive sanitary picture of the city. The most famous of these surveys was the *Report of the Council of Hygiene and Public Health of the Citizens' Association of New York upon the Sanitary Condition of the City* (1866), which would a play key role in securing a permanent board of health in New York City.

Sanitary surveys provided a "ground level" view of local conditions. In a sense they rendered the city legible, as they were the primary instrument by which sanitarians transformed sensory observations into first, systematic data and second, visual dot maps that spatialized the data of disease. Such a "thorough and systematic sanitary inspection by competent experts" (Citizens' Association of New York 1866, xxii) would confirm:

> what reason should have taught every person, however uneducated, that filth, overcrowding, bad drainage, excessive humidity, imperfect supply of air and sunlight, neglect of excrementitious and decaying material, and the putrid exhalations from sinks, sewers, gutters, and dirty streets, will both produce and perpetuate disease; and that whatever sickness occurs in such localities will be more virulent and destructive than the same or similar maladies when occurring in places where such conditions do not prevail. (Citizens' Association of New York 1866, lxiii)

Through the sanitary surveys, local knowledge and anecdotal observations were systematically amassed so as to transform them into the generalized, abstract representations of dot maps and normal morality ratios.

The combined effect of dot maps, normal mortality ratios, and sanitary surveys was to produce an understanding of cholera as filth, linking the disease to place in an understanding that was inherently ecological. This was no more evident than in their identification of "cholera nests." Cholera nests, also referred to as "cholera fields" (*New York Times* July 19, 1866, 2), "diarrheal fields" (Metropolitan Board of Health, 1867: 150) , "fever nests" (Citizens' Association of New York 1866, 35), and "plague spots" (*New York Times* July 19, 1866, 2), were those "special centres" (Metropolitan Board of Health 1866, 38) that possessed multiple factors "from whence emanate the most dreaded diseases that find their way to the more favored districts of the city" (Citizens' Association of New York 1866, xxxiv). The Metropolitan Board of Health (1866, 13) described one of these cholera nests:

> The streets were uncleaned; manure heaps, containing thousands of tons, occupied piers and vacant lots; sewers were obstructed; houses were crowded, and badly ventilated, and lighted; privies were unconnected with the sewers, and overflowing; stables and yards were filled with stagnant water, and many dark and damp cellars were inhabited. The streets were obstructed, and the wharves and piers were filthy and dangerous from dilapidation; cattle were driven through the streets at all hours of the day in

> large numbers, and endangered the lives of the people; slaughter-houses were open to the streets, and were offensive with accumulated offal and blood, or filled the sewers with decomposing animal substances. Gas companies, shell-burners, and fat-boilers, pursued their occupations without regard to the public health or comfort, and filled the air with disgusting odors.

Cholera's "favorite haunts" (Newman 1856, 441) were breeding grounds for all types of disease and, as such, were a menace, not only to their inhabitants but also to the rest of the city. Because cholera "will be *epidemically propagated* only where and when certain conditions of putrescence in the earth, the atmosphere, or the potable water are present" (Metropolitan Board of Health 1867, 150), the areas in which these "localizing conditions of disease" (Citizens' Association of New York 1866, lxiii) were rife needed to be cleaned to prevent an epidemic. As cholera became rooted in place, those responsible for filthy places (i.e., landlords) were put on watch as culpable purveyors of disease.

The relationship between cholera and filth remained vague and, therefore, could accommodate divergent viewpoints regarding the etiology of diease. On the surface, it seemed to confirm the views of noncontagionists at the expense of contagionists. After all, it focused on local conditions in accounting for disease. Many historians have taken this as a sign that noncontagionism was dominant (e.g., Rosen 1993). But while it is true that noncontagionism was gaining adherents, the idea of cholera as filth had more ambiguous effects. The flexibility in the understanding of the relationship between filth and cholera enabled contagionists and noncontagionists, homeopaths and regulars, to unite in the common goal of cleaning the city. Contagionists bought into sanitation by viewing the local conditions essentially as intervening variables in the spread of the disease, a common position during the period called "contingent contagionism," which held that cholera needed a ripe environment to prosper and grow (Hamlin 2009). Local factors were necessary but not sufficient for the spread of cholera. As for strict contagionists, the focus on sanitation could not hurt, provided it did not detract from quarantines and the search for the etiological cause of cholera.

Even diverse medical sects converged on a common ground of cholera as filth, as elements of the argument resonated with their different epistemological systems. Homeopaths, long amenable to the type of statistical rhetoric being proffered by sanitary reformers, were convinced of the association between filth and disease. In a speech on cholera before the American Institute of Homeopathy, Horace Paine (1866, 142–143) outlined the homeopathic position:

> Whatever opinion may be entertained as to the cause of cholera or the mode of its propagation (and on these points physicians are still much divided), it is certain that foul and confined air, putrid and decaying vegetable and animal deposits, and damp, crowded, and ill-ventilated apartments, offer the greatest encouragement to its development and increase its virulence. Consequently, cleanliness of towns, dwellings, and persons, is of the first importance; and would, if fully accomplished, be more effectual than any quarantine arrangements in preventing its invasion.

For their part, regulars could support the link between cholera and filth, even though their persistent suspicion of statistics made them unlikely to accept the normal mortality ratio. The link conformed to their clinical observations of cholera's tendency to cluster in certain neighborhoods. And the shoe-leather data of the sanitary surveys resonated with allopathy's commitment to radical empiricism and sensory observation. Provided sanitarians' claims remained rooted in these observations, regulars were willing to sign on.

Therefore, while medical debates about the nature of cholera endured, they were mitigated by a growing focus on prevention. Elisha Harris (1869, 122), a longtime sanitary reformer and commissioner for the national Sanitary Commission, reporting to the New York Academy of Medicine in 1866, stressed "that physicians and sanitary authorities should promptly act upon all practical questions relating to this duty *without waiting the adjustment of theories*, remembering that it is the first duty to prevent or control the earliest outbreaks of the disease" (emphasis added). Because the association of cholera with filth did not require the commitment to a *causal* understanding of the relationship, it united strange bedfellows. As the *New York Times* (July 1, 1866, 4 emphasis added) pointed out in discussing the New York Academy's support of sanitary reforms in New York City,

> The camps of the medical world are divided on this disease, as on many others, into two hostile parties, for and against contagion; but the truly scientific, like the members of the Academy, are waiting patiently for facts

> before they proclaim a theory. Thus far, in the approach of this epidemic, the facts are sufficiently confusing. . . . *The Academy, wisely, do not commit themselves for or against any theory of contagion; they only urge the practical course of cleansing and disinfection as "absolutely necessary."*

The ambiguity and flexibility in the understanding of the relationship between filth and disease made it hard to reject whatever one's particular theory of disease. All could agree on some generic sanitary reforms.

INTELLECTUAL ECUMENISM AND EPISTEMIC TRUSTWORTHINESS

The localization of cholera necessitated locational responses. Unified under a broad banner of sanitary science, reformers of diverse backgrounds formed a coalition to advocate for the cleaning up of the city. This coalition sought to harness a wide array of technical knowledge and community energy to conquer disease by creating permanent, independent boards of health. But, in doing so, they faced stiff resistance from local political machines. The existing sanitary regimes were bastions of corruption and political patronage. To bring about the independent boards, sanitarians had to convince legislatures of their superiority over the existing boards. They had to convince legislatures that they possessed better knowledge and, in turn, better protection against cholera.

To justify their epistemic authority, sanitarians adopted a position rooted in their disinterested position and the accompanying trustworthiness of the knowledge that emanated from such a position. This strategy was born of the particular configuration of the reform movement. The diverse knowledge that sanitarians valued precluded them from claiming epistemic authority in terms of possession of any one type of specialized knowledge. Nor could the sanitarians claims special technical acumen in producing dot maps, statistical data, and sanitary surveys, for politicians used similar techniques to make contradictory arguments. Rather, they argued that existing boards of health and sanitary policies of the city were unsalvageable because of political corruption. Sanitary contracts were "managed in the interest of partisan greed" (Agnew 1874, 4), doled out as patronage by urban political machines in exchange for political support. Streets went uncleaned, inspections unconducted, and water unpurified, all because city inhabitants "stupidly permit[ted] the conditions of their health to be controlled by politicians" (*Nation* January 11, 1866, 40). Politicians were not just ignorant; they lacked the proper motivation to honestly carry out their duties. "The doctrine that 'to the victors belong the spoils' may be food for politicians and a fair rule of practice in the general affairs of State," a sanitarian later echoed, "but when positions of public trust are to be filled whose duty requires technical ability and special adaptation, then it becomes highly expedient that politicians be requested to keep hands off" (Halley 1887, 241). Reduced to political transactions, sanitary activities were not taken seriously, nor was there any incentive to produce accurate knowledge of sanitary conditions.

Instead, sanitarians stressed the disinterested and apolitical nature of their motivations. Sanitarians were uncorrupted by political calculations and driven by purer motives of civic responsibility. They wanted to see the streets cleaned and wanted to figure out the best means to do so. Their knowledge was unsullied by the kickbacks and graft that accompanied machine politics. Thus, it was disinterested knowledge, not technical acumen, which they offered as the antidote to political corruption, local filth, and epidemic disease. In other words, sanitarians sought to win back public trust through a particular ethos of disinterestedness toward sanitary knowledge. This ethos was joined with new technologies—accurate cholera maps, comprehensive sanitary surveys, and troubling mortality statistics—that sought to render knowledge of disease visible and transparent to the public. Transparent sanitary knowledge was opposed to the backroom dealings of politicians. It promised both more accountability and better accuracy.

Sanitarians argued that boards of health needed to be insulated from politics, staffed by officers who would not abuse the public trust for political gain and who would apply the proper reforms based on trustworthy knowledge. This argument was not a claim to objectivity per se, as it is commonly understood today. It did not rest on an ideal of aperspectival knowledge, a "view from nowhere," which argues for privilege based on having eliminated (or at the very least, greatly minimized) potential biases and corrupting influences of particular perspectives (Novick 1988)—a claim to trust based on a negation. Sanitarians had no pretensions of purging the positionality of the knower from the production of knowledge. To the contrary, positionality and the *character* of the knower determined the epistemic validity of knowledge claims. Sanitarians' trustworthiness was based on their standing as responsible citizens in the community unbiased by crass politics. Their epistemic authority was inextricably tied to who they were and where they stood vis-à-vis city politics. This was an argument for epistemic authority that was, in an important sense, ad hominem. By virtue of who they were sanitarians could produce accurate and honest sanitary knowledge that would rid the city of filth.

The First Permanent Board of Health

The struggle for sanitary reform in New York City offers insights into the way in which reformers advocated for and won important reforms by framing their epistemic authority in terms of disinterested, apolitical knowledge. Sanitarians in New York not only secured the first permanent, depoliticized municipal board of health in the United States; their efforts became the model for sanitary reformers in other contexts.

New York, like many other U.S. cities, had established temporary boards of health during prior epidemics to disseminate information and coordinate measures to prevent the spread of the epidemic. These temporary boards typically consisted of little more than the mayor and alderman, with some input from a few well-regarded physicians who sat on hastily thrown together special medical councils. There would be a period of porous quarantines and cosmetic cleanings, but once the epidemics passed, the sanitary situation would return to its pre-epidemic condition. Lingering sanitary practices, like street cleaning, were doled out as patronage to supporters by city political machines, resulting in a hodgepodge of municipal organizations working independently of one another, more interested in advancing their political agendas than cleaning the streets (Duffy 1990).

The initial call for reforming New York's municipal health was led by the allopathic medical profession itself. While rank-and-file regulars tended to be ambivalent toward public health, a handful of elite physicians, driven by a sense of civic duty, prodded the New York Academy of Medicine (NYAM) into advocating for a more permanent solution to sanitation in 1857. They supported legislation that would depoliticize sanitation. Within allopathy, this spurred internal debate as to the relevance of public health, as well as concerns that putting energy into sanitary reforms was misguided. Still, even the skeptical rank and file, less encumbered by civic duty, could see the potential professional benefits of reform, provided that public health remained under allopathic control. The president of the Medical Society of the State of New York, Fordyce Barker (1860, 8), made the case for allopathic control:

> And I say further, that it is the absolute duty of the citizen, and the duty of the legislator, to accept as final our decisions in all scientific hygienic matters, and to act upon such decisions. As the decisions of a court of last resort furnish a final interpretation of written and shape the unwritten law for the legislator and the citizen, so should the expressed convictions of this scientific body be received and accepted by all, and be the basis of all action had in sanitary matters. Knowledge on this subject can only be had from medical men.

In arguing for allopathic control, sanitary-minded physicians aligned the calls for reform with the professional goals of the society, and thereby gained NYAM's endorsement.

These early reforms received a boost from newspaper exposés that blamed the poor sanitary condition of the city on political corruption. The *New York Times* (June 25, 1856, 3), in an editorial provocatively entitled "Killing Off Our Children—By Authority," drew the link between patronage and needless death: "Thousands of lives are lost, and thousands that live are demoralized and broken in constitution—all because our mock sanitary officers have not the intelligence or the enterprise to fill up these sunken lots, and drain these poisoned valleys." The not-so-subtle moral of this editorial was clear: the greed of politicians was responsible for the fever nests that killed children. Exposés like this focused public attention on the harms caused by lax sanitation and placed culpability for filth squarely on the shoulders of unresponsive city officials.

NYAM, however, was not able to translate this support into tangible reforms. Its plans were derailed by the city's machine politics, as the Democratic political machine of Tammany Hall persuaded the state assembly to reject the proposed legislation. The legislative failure gave those within the allopathic profession initially skeptical of reforms fodder for backing out of future reform efforts. They dismissed any further forays into public health reform as a waste of time and resources, given their political unlikelihood and marginal payoff. Consequently, after this brief flirtation with sanitary reform, NYAM, and regulars in general, remained content to sit on the sidelines as others took up the fight against entrenched political interests (Duffy 1968). Physicians were free to support the cause individually, but without the support from the medical society.

When the academy bowed out, the reform mantle was assumed by a diverse group of sanitarians. Whereas the regulars proposed reforms focused on medical control, the subsequent push for reform offered a more ecumenical vision of sanitary science. Elite, sanitary-minded physicians, like John Griscom, Elisha Harris, Joseph Smith, and Stephen Smith, collaborated with community leaders, like wealthy industrialist/philanthropist Peter Cooper. This eclectic group of elites was united by a commitment to responsible citizenry, not by any notion of professional expertise, for while physicians were central to the burgeoning sanitary movement, the bulk of the movement consisted of lay community members. As such, calls for sanitary reform took on a less explicitly medical character. First through the New York Sanitary Association and later the Citizens' Association of New York, sanitarians worked to portray the existing boards as woefully and irredeemably corrupt. They produced reports and pamphlets that connected disease to filth and filth to political corruption. Their justification for special recognition rested on the purported apolitical disinterestedness of their sanitary science. "Politicians make poor sanitarians," because it was against their political interests to produce accurate sanitary knowledge requisite for effective intervention (Halley 1887, 241). From 1859 to 1866 reformers continuously introduced sanitary bills to the legislature, only to be defeated by the Democratic Party machine and "paltry officials who hang like leeches to the municipal body," as the *New York Times* described opponents of the bill (Duffy 1968, 546). Defeating such entrenched interests was proving to be a heavy lift.

In 1863, the Citizens' Association established the Council of Hygiene and Public Health, charged with conducting a thorough sanitary survey of the entire city. The new survey was carried out in order to rebut a city inspector's report, which denied the existence of any sanitary problems. The comprehensive *Report of the Council of Hygiene and Public Health of the Citizens' Association of New York upon the Sanitary Condition of the City* was an impressive feat of investigation. The council divided the city into thirty-one specified districts, assigned an inspector to each district, and supplied them with a standardized form to record the sanitary conditions of their district. Inspectors were instructed to gather information on a wide variety of conditions, from "the nature of the ground" to the "location and character of water closets" in tenement houses (Smith 1911, 54–55). Stephen Smith (1911, 54–55), author of the final report, outlined the exhaustive list of relevant factors that investigators were asked to note:

> Commencing at a given corner of the district, he [the investigator] was first to go around the square and note: 1. Nature of the ground. 2. Drain-

> age and sewerage. 3. Number of houses in the square. 4. Vacant lots and their sanitary condition. 5. Courts and alleys. 6. Rear buildings. 7. Number of tenement houses. 11. [*sic*] Drinking shops, brothels, gambling saloons, etc. 12. Stores and markets. 13. Factories, schools, crowded buildings. 14. Slaughter-houses (describe particularly). 15. Bone and offal nuisances. 16. Stables, etc. 17. Church and school edifices.
>
> Returning to the point of starting, he was to commence a detailed inspection of each building, noting: *a*. Condition and material of buildings. *b*. Number of stories and their height. *c*. number of families intended to be accommodated, and space allotted to each. *d*. Water supply and house drainage. *e*. Location and character of water closets. *f*. Disposal of garbage and house slops. *g* Ventilation, external and internal. *h*. Cellars and basements, and their population. *i*. Conditions of halls and passages. *j*. Frontage on street, court, alley—N.E.S. or W.18. Prevailing character of the population. 19. Prevailing sickness and mortality. 20. Sources of preventable disease and mortality. 21. Condition of streets and pavements. 22. Miscellaneous information.

The identification of so many different conditions demonstrates the breadth of sanitary knowledge. When aggregated into a single report, the findings were both surprising and disturbing, as the report revealed the presence of cholera nests and unearthed an invisible smallpox epidemic to boot. The sheer scale and detail of the *Report* impressed the media and the public, who joined in denouncing the sanitary situation of the city. By making disease in the city legible, sanitarians had produced a comprehensive condemnation of politicized sanitation (or lack thereof).

In addition to the *Report*'s revelations about the failure of city officials to perform their trusted duties, additional investigations found outright fraud. In conjunction with the survey, the Citizens' Association issued a pamphlet chronicling the abuses of the City Inspector's Department and showed that despite a generous allocation of funds for sanitation, little of the money actually went to cleaning up the city: "Under the present rule of ignorant and corrupt politicians, this city expends directly and indirectly nearly half a million of dollars for health purposes, not *one dollar* of which is intelligently applied to improve its sanitary state. Small-pox, scarlet fever, cholera infantum, and allied diseases, rage among the poor like consuming epidemics without one effort put forth by our *one hundred and eighty-three* health officials" (*American Medical Times* 1862, 99). Politicians were not just inept;

they were corrupt. They had abused their power and betrayed the public's trust.

With the reports attracting widespread attention, reformers reintroduced a bill in 1865, hoping to capitalize on the renewed outrage. This time reformers received a much-needed boost from an unlikely ally—cholera. The arrival of cholera in port aboard the *Atalanta* in November brought a new sense of urgency to the proposed reforms. Although cholera remained contained to the port until spring, the looming specter of another epidemic lent credence to the reformers' calls. The disease was poised to attack the filthy city again. Faced with a looming epidemic, the press ratcheted up their critiques of the City Inspector's Department and their editorial support for the reforms. While *Harper's* mocked the board's apathy in a cartoon, the *Nation* ("Street Commissioners to the Cholera" November 9, 1865, 583) did so in verse:

> Cholera, cholera, cholera, come!
> Come to the city we dock for thy home! Come to Manhattan!
> New York never gave
> Prince, hero, charlatan, exile, or knave,
> Cholera, such a reception as we,
> Queen of men's terror! Have plotted for thee!

By not attending to the egregious filth, the city was rolling out a welcome mat for cholera, so went the criticisms.

Pressured by sanitarians, the media, and the looming threat of cholera, the legislature passed "An Act to Create a Metropolitan Sanitary District and Board of Health Therein." It became law on February 26, 1866. The sanitarians' dream of an independent board of health was finally realized. The decade-long reform effort came to an end just as cholera arrived.

The composition of the first permanent board of health and its wide-reaching activities indicated both a particular understanding of cholera—cholera rooted in place as filth—and a broad view of relevant knowledge. Sanitation was framed as thoroughly social, not narrowly medical. While physicians were recognized as an integral part of the Metropolitan Board of Health of New York City, the legislation expressly ensured that they would remain in the minority, as reformers did not want the board to be dominated by doctors (Duffy 1968, 2). The board consisted of four police commissioners, the health officer, and four other commissioners appointed by the

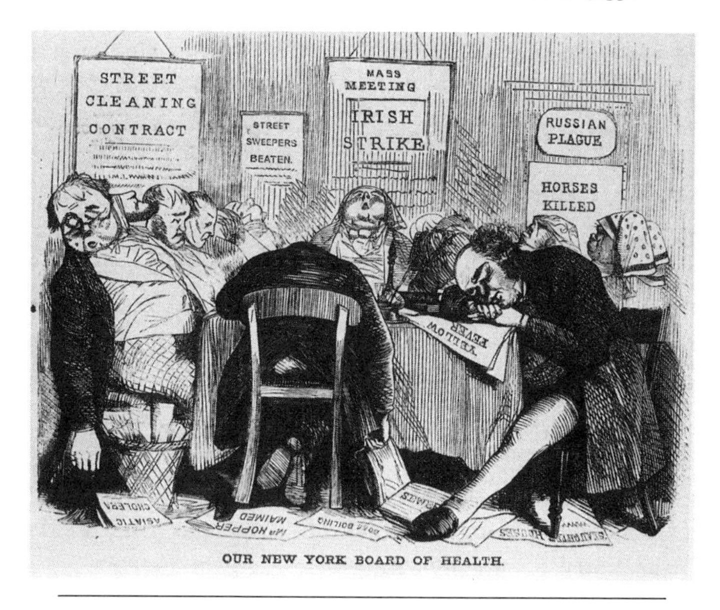

Cartoon showing political board appointees asleep on the job from *Harper's Weekly*, August 5, 1865, p. 496. Courtesy of the National Library of Medicine.

governor, three of whom had to be physicians. Valued as having pertinent insight, physicians had a total of four seats on the board, but lay community members maintained the majority with five, and leadership positions went to nonmedical members. Thus, the board was decidedly *not* the medical organization under the control of allopathic physicians envisioned by NYAM in 1857. Rather, it was constructed as a coalition of lay community members and relevant professionals committed to sanitary science as a cooperative endeavor patterned on "mutually supporting roles" (Rosenkrantz 1974, 58). This reflected the intellectual ecumenism of sanitary science and the cooperative, but not dominant, place of medical knowledge within it. The broad understanding of the relationship between disease and the environment precluded defining sanitary science in terms of medical knowledge only.

The actions of the board reflected the numerous forms of knowledge it brought to bear on cholera. When Governor Fenton issued a "Proclamation of Peril" on April 24, giving the board wide authority to carry out the reforms necessary to combat cholera, it immediately set to work (Duffy 1974). Two thousand police officers fanned out across the city, reporting every instance of sanitary neglect over the fifty telegraph lines that the board had established throughout the city (Chambers 1938, 274). Upon notice, the board removed the offending nuisances, as it took control of the street-cleaning contracts and garbage removal. In dealing with cases of cholera, doctors were instructed to report all cases to the board, which would confirm each case and quickly remove those with cholera to quarantine hospitals. Cholera nests were destroyed, their inhabitants relocated, while salvageable buildings were disinfected. Furthermore, the board closed the most egregious of the city's polluters (e.g., fat- and bone-boiling establishments), sent engineers to investigate new buildings for adequate ventilation, and organized a corps of physicians to staff cholera hospitals. As they went well beyond medical interventions, these diverse actions required the input and coordination of a wide array of experts.

The board's proactive approach seemed to pay off. Cholera threatened the city throughout the summer of 1866 but never really gained a foothold. Whether legitimate or not, the board received credit for the epidemic's mildness (Duffy 1968, 18). In its review of the epidemic on its one-year anniversary, the *New York Times* (March 31, 1867, 3) applauded the board:

> The community was watching with considerable anxiety for the advent of cholera, and speculations were freely indulged as to the effect which its presence would exert upon the health and business of the City. A newly-formed Board of Health had entered upon its duties; people were hoping that it might in some way be instrumental in averting or diminishing the ravages of what had been in previous years a terrible scourge. Although it may be said that no inference can be drawn from the fact that mortality from cholera last year in New-York was less than during the previous visitation, we are unwilling to admit the validity of the assertion. In other cities on the continent where the disease obtained a foothold the loss of life was fearful, and certainly the conditions which favored its spread existed in full force in New-York. Hence we believe that the methods adopted by the Board of Health were instrumental in saving many lives and even in limiting the epidemic.

Accepting the *Times*' evaluation, other cities followed suit and established their own municipal boards of health, as the Metropolitan Board of Health became the model for sanitary reformers throughout the country (Duffy 1990; Rosenberg 1987b). In these cities, the sanitary reforms followed a similar trajectory; spurred by the threat of cholera, a coalition of politicians, physicians, and social reformers agitated for and won reform. State boards of health were also formed, beginning with Massachusetts in 1869. Even a short-lived National Board of Health was created in 1879 (Smillie 1943). As reforms spread, the sanitary movement, formerly a hodgepodge of reforming physicians and concerned citizens, became organized. A collective identity crystallized among sanitarians and was formalized with the founding of the American Public Health Association (APHA) in 1872. Unlike the AMA, the APHA was not a professional association but "a body of informed persons of good will to facilitate the enlightenment of the public and promote the appointment of more competent health authorities" (Smith quoted in Rosenkrantz 1974, 58). It became the central node for a wide array of diverse actors in the growing national movement, facilitating exchanges between diverse local reformers. No longer isolated in particular communities, to be a sanitarian now meant belonging to a national community of reformers, a diverse community, but a community nonetheless. By the early 1870s, the public health movement was nationally popular and institutionalized.

A COOPERATIVE SUCCESS, A PROFESSIONAL PROBLEM

With their popularity and resources, boards of health became a key prize in the epistemic contest. One might surmise that given the initial support regulars displayed toward sanitary reform and the participation of elite, sanitary-minded physicians on the board that the regulars would view the boards as a positive occurrence, an opportunity to advance their professional and epistemic goals. This was not the case. To the chagrin of regulars, the public health movement was inclusive and ecumenical in nature. This framing, which had allowed sanitarians to overcome entrenched politicians, created unintended problems for allopathic physicians in their epistemic contest over disease with homeopaths and other sanitarians that hindered the professional agenda of the AMA and local allopathic societies.

Despite the widespread celebration of public health after the 1866 epidemic, the bulk of allopathic physicians developed ambivalence toward the boards and sanitary science in general. On one hand, the boards of health seemed to offer an opportunity to improve the public image of allopathy (Rosenkrantz 1974). In 1873, President C. R. Agnew of the Medical Society of the State of New York urged his colleagues to embrace public health in order to gain "a new and enduring title to the respect and the gratitude of the public" (Agnew 1874, 4). Sanitary-minded physicians appealed to nobler sentiments, pointing out that "the vocation of the medical man is not bounded by the narrow confines of curing the sick, but embraces a far nobler work—a work of illimitable extent—the prevention of disease, and the prolongation of life; a field of science 'where the harvest is truly plenteous, but the laborers are few' " (*American Medical Times* 1860, 47). The *American Medical Times* (1860, 46) argued that to "defend and relieve our fellow men from the preventible [*sic*] causes of disease, is manifestly the highest mission and best service of medical science and skill." According to sanitary-minded physicians, this "highest mission" of the profession required doctors to put aside professional concerns and work cooperatively to end disease and, in turn, put themselves out of business.

Not all allopathic physicians agreed with this noble sentiment. Cholera as filth demanded wide-scale reforms, ranging from the disinfection of tenement cellars to new ventilation systems, from water sanitation to street cleaning. Reformers lacked any means by which to weight the impact or relevance of one factor vis-à-vis any other. The embrace of public health, therefore, required the embrace of a nonhierarchical, cooperative spirit, precisely at the time during which regulars sought to distinguish themselves as possessing epistemic authority. Because "cholera should not be treated as a disease, but as a pestilence" (Smith 1869, 59), sanitarians embraced an ethos of intellectual ecumenism anathema to the professional agenda of the AMA, embodied in the no consultation clause. Regulars not directly involved in the sanitary movement balked at this ecumenism, for it inhibited the profession's claim of a privileged standing within sanitary science and precluded their attempts to control the definition of cholera. Medical knowledge was placed on par with other forms of knowledge.

This was all the more troubling given the demands the boards of health asked of physicians—they wanted physicians to forfeit some of their autonomy by reporting their own cases to the boards and turning over their cholera patients to board-controlled hospitals. Most rank-and-file regulars did not welcome nonmedical board members inserting themselves between them and their patients. Their dismay was manifest in their widespread hesitation, and often blatant uncooperativeness, in reporting cases of disease to the board (Hammonds 1999; Maulitz 1979). They feared that their patients would be taken away from them, removed to one of the cholera hospitals to be treated by another doctor or, even worse, a homeopath. In this way, the boards' demands were contradictory to the economic logic of allopathic medical practice, abhorrent to their self-interest. There was a class dimension to allopathic ambivalence toward public health. Most of the sanitary-minded physicians were wealthy elites from families with a historical commitment to moral reform and civil service. While their medical knowledge was valued in the sanitary movement, their participation was rooted in their standing as important members of the community, "less in scientific acumen than in responsible citizenship" (Rosenkrantz 1974, 58). Rank-and-file regulars felt that they could ill afford the luxury of participation, given their precarious professional and economic situation. Indeed, the AMA's professionalization strategy was to deny the legitimacy of control over medicine by anyone other than allopathic physicians themselves. The entire premise of their professional project was to gain control over disease, not to share it with others. And this is precisely what the sanitary movement was asking them to do. Therefore, while the elite, sanitary-minded physicians preached cooperation, the majority of rank-and-file regulars felt that the professional benefits of sanitary science could only be accrued if and when allopathy controlled the boards of health.

Professionalizing Plumbers?

In addition to their troublesome economic implications, the boards of health encouraged even more challenges to allopathic authority, as a variety of new actors began to assert their right to participate in the management of disease. By rooting disease in place, the idea of cholera as filth *extended* the contours of the epistemic contest beyond the bounds of medicine. Other "experts" had something relevant to say. This expansion was exemplified by the attempt of plumbers to improve their own standing through their association with public health. Prior to 1866, it would have been unthinkable for allopathic physicians to envision a challenge to their authority from plumbers. After 1866, it was the reality of the epistemic contest.

Like the sanitary reformers generally, plumbers justified sanitary reforms on apolitical grounds: "In order that it [the board of health] may be of benefit to the people and extend its usefulness to its full capacity, its complexion should not be characterized by any sort of partisanship. It should be composed of sanitarians, if the objects of the health board are to be attained" (Halley 1887, 241). Plumbers could better serve the people's trust because they had the requisite expertise and were immune to political corruption. And because sewers, cesspools, and inadequate internal plumbing contributed to disease, plumbers possessed valuable insights into the transmission of disease and the creation of cholera nests. As the "water supply of cities and dwellings, the sewerage of cities and house drainage, are some of the most important features of sanitary science" (Halley 1887, 243), plumbers' knowledge was needed to unearth the mystery of disease. And unlike the doctor, the plumber had both the skill and motivation to search out disease in its element:

> No doubt he [the doctor] often speculates as to what may have produced the trouble he is striving to remedy, but when it comes to a careful, minute, scientific investigation to determine this cause, he usually has neither sufficient motive, time nor knowledge to make it. Is it a matter connected with defective sewage in relation to—suppose we say diphtheria? This involves questions of sanitary engineering, the work of the plumber, the composition of sewer gases, and tests for them. (Billings 1879, 125)

Possessing this requisite expertise, plumbers demanded their due recognition within public health, to argue that "we are not arbitrary" (Halley 1887, 245) when it came to sanitary science. William Halley (1887, 245), a master plumber from Ohio, drew on the link between disease and place to demand inclusion on the boards:

> No one profession or calling is able to hold sanitary science in the hollow of its hand. It has been the impression among legislators that physicians alone are competent to serve as members of State boards of health. Not all physicians are sanitarians, and the best medical ability on earth is unable to compass the whole range of *practical* sanitation. It is granted that medical knowledge is indispensable to a health board, but it requires the learning and skill of other callings to complete the full complement of a thorough, practical, efficient sanitary organization. Sanitary science to be of value must be practical; and the varied aspects of sanitation cannot be reached by medical wisdom alone, but by the technical knowledge and skill of mechanical art.

As paragons of "practical sanitation," plumbers demanded not only inclusion but also the authority to carry out their work autonomously, unquestioned by other, nonplumbing sanitarians.

While sanitary-minded physicians attested to plumbers' relevant sanitary knowledge, the bulk of allopaths ridiculed their aspirations. For regulars, cholera was a medical problem, and when dealing with disease, it seemed absurd that plumbers would dictate anything to physicians. T. Clarke Miller (1887, 109), regular physician and president of the Ohio State Board of Health, mocked the plumbers' attempt to assert themselves on the boards:

> The plumbers have been making very commendable progress in the past few years; there are many of them who can be trusted to do safe work in their line; but some of them, by reason of having learned their trade well, have come to consider themselves commanders in the sanitary army, and to assume that the whole burden of practical sanitation rests upon their shoulders, and to felicitate themselves that doctors and architects have heartily joined with them, who have always done, and are still doing, everything that is necessary to do about the relations of plumbing to health. Physicians who kept themselves abreast with the advance of knowledge of the etiology of disease, and have been foremost in the detection of bad plumbing by a knowledge of its results, are not only unnecessary but meddlesome—they know how to cure disease—nothing about how to *prevent* it. . . . Indeed, we are in danger of having an army of trades seeking to place themselves in the sanitary priesthood. . . . The butcher who knows wholesome meat and furnishes it to his customers is likely to become the "sanitary butcher."

Sarcasm aside, Miller was serious in his foreboding. The medical profession could not afford to ignore "the growing army of sanitarians in special lines" (Miller 1887, 108). Regulars might lose their uniqueness and distinction within the sanitary movement if every group professionalized in the name of public health.

The boards created an arena accessible enough, epistemologically speaking, to allow for challenges to allopathy that extended beyond alternative medical sects; plumbers and other "sanitary professions" entered the fray. The added competition complicated regulars' attempts to use the boards to promote their professional ends. This was not just frustrating; it was nearly incomprehensible. Regulars believed that medical knowledge should trump other forms of knowledge, for "it will be seen that one may be a good practicing physician without being a sanitarian, but no one can be a good sanitarian without being a good physician; the requisites for one underlie those for the other, as a foundation" (Griscom 1857, 110). Insofar as nonmedical knowledge was relevant, allopathy sought to relegate such knowledge to a secondary status and to assert its control over it. Either physicians should dictate how the complementary, but inferior, knowledge should be deployed or physicians should acquaint themselves with this knowledge and deploy it themselves. Thus, although allopathic physicians begrudgingly acknowledged the relevance of other knowledge, they sought to subordinate it and, in some cases, even to steal it. Allopathic physicians envisioned a system of public health to be managed by doctors and doctors only.

To the extent that plumbers were never able to gain equal standing to physicians on the boards, allopathic physicians were successful. In most matters, medical knowledge remained valued over knowledge of plumbing. Unsurprisingly, plumbers bristled under this situation: "Last and worse of all, in spite of these severe labors, the plumber is subject to general obloquy. No one appreciates his labors. His bills are usually disputed, and even when he has done his best he feels that it is a thankless task and that no one appreciates his efforts" (*Plumber and Sanitary Engineer* 1879, 94). Still, plumbers were granted a place at the public health table as active members of boards of health. They were successful in their campaigns against irrelevancy and were able to prevent doctors from fully capturing their knowledge, retaining some autonomy from sanitary physicians. These efforts—along with similar efforts on the part of civic engineers, architects, and other sanitarians—ended up tainting the original promise of sanitary science for allopathy, as regulars were finding it more and more difficult to assert themselves among the cacophony of competing voices.

Homeopathic Incursions

Arguments for allopathic control over the boards faltered in the face of the recognition of the diverse components relevant to public health. But even more troubling for regulars was their failure to maintain control over *medical* knowledge. The establishment boards not only encouraged other sanitarians to challenge allopathy; it emboldened the homeopathic challenge, as homeopaths increasingly demanded equal status to regulars. Furthermore, the manner in which the boards were justified epistemologically—as producing apolitical knowledge—constrained regulars' attempts to exclude other medical sects from participation. In this favorable context, homeopaths jockeyed for, and typically won, inclusion on the boards. More than anything else, the inclusion of homeopathy aggravated rank-and-file regulars.

Take the experience of the New York Board of Health. Initially, when the Metropolitan Health Bill was being debated in the legislature, homeopaths petitioned for inclusion. Anticipating an allopathic effort to exclude them, the Homeopathic Medical Society of the State of New York (1866b, 131–132) warned the legislature that a "great injustice would be done if a provision is not made for the appointment of at least one of the medical men composing the board from the ranks of the homeopathic school; and also, if at least, one fourth part of all the public hospitals and dispensaries in the city of New York, should not be placed in the care of homeopathic physicians." The legislature was in a bind. Politically, it agreed with homeopaths, but it could not risk alienating allopaths. Rather than addressing such a contentious issue, the legislature punted, remaining silent and declining to make any specifications as to which sect should be represented in the appointments. Once the bill passed, however, the governor appointed four regular physicians who had been active in the sanitary movement. Although this group was less explicitly engaged in the professionalization project of the AMA and NYAM, regulars celebrated the exclusion of homeopaths from the board.

The enthusiasm was short-lived. As the board made preparations for cholera, homeopathic physicians lobbied for control of at least one cholera hospital. Many homeopaths wanted to use the cholera epidemic as an opportunity "to show the superiority" of their system (Homeopathic Medical Society of the State of New York 1866a, 247). Homeopaths believed that the existence of a homeopathic cholera hospital would create a real-life experiment, whereby the treatment outcomes of the homeopathic hospital could be compared to those under allopathic control. Appealing to past statistical evidence, homeopaths were confident that their system would prove to be more effective—"In no disease has the value of our treatment been more satisfactorily shown, than in epidemic cholera, the statistics of which have been frequently published" (Homeopathic Medical Society of the State of New York 1866c, 322)—and sought to make use of the epidemic to put to rest allopathy's arguments that homeopathy was quackery and need not be taken seriously.

Rather than merely requesting recognition, homeopaths *demanded* it "as a right, that so large a portion of our tax-payers should be fairly represented in our medical institutions" (Homeopathic Medical Society of the State of New York 1866c, 322). E. A. Munger (1865, 22), president of the Homeopathic Medical Society of the State of New York, argued that homeopaths were "entitled to all rights and privileges enjoyed by allopathy." This position was grounded not only in homeopathy's equal legal standing, won in the repeals of medical licensing laws, but also in the very rationale of the board. Homeopaths adopted the same trope used by sanitarians generally: adequate sanitary knowledge had to be, by necessity, apolitical and disinterested. Base political motives, professional or otherwise, corrupted knowledge and hampered sanitation. Thus, homeopaths denounced all allopathic arguments as political; exclusion from public health was equated with political "oppression" (Cornell 1868, 4). By painting allopathic physicians with the same political brush that sanitarians painted corrupt politicians, homeopaths offered a rationale for their inclusion that mirrored the rationale of the boards more generally. Underlying this was the familiar homeopathic appeal to democracy that had been successful in the past. Dialogue, transparency, and participation would lead to better knowledge, and, in turn, better sanitation. In demanding inclusion, homeopaths were only asking to "let the public judge of the candor, intelligence, and honesty, with which the claims of homeopathy are treated by men from whom we have a right to demand justice" (Bowers 1868, 406–407).

Allopaths also rehashed old arguments. As in the debates over licensing, regulars balked at the suggestion that homeopaths should receive official government recognition. They adamantly opposed homeopaths treating *any* cholera patients, much less an entire hospital's worth. Many rank-and-file regulars derisively mocked homeopaths' interest in public health, facetiously wondering how their commitment to infinitesimal dosages would translate into public health interventions: "We have never known such remedies applied there for cleaning the streets, the removal of nuisances, the repair of defective sewage, except in the most diluted *infinitesimal* doses. What is needed are not little homeopathic wheelbarrows, but big, 'allopathic' four-horse teams, to remove the dirt and filth from New York City" (*Medical and Surgical Reporter* 1866b, 95–96). Having defined homeopathic knowledge prima facie as quackery, regulars felt no compunction to engage in "wanton experimenting" (Blatchford 1852, 136) with homeopaths. The *Medical and Surgical Reporter* (1866a, 477) warned that if homeopaths had the "right to ask that the public charities shall be made to afford to them an opportunity to experiment with the poor," then "the Metropolitan Board of Health would soon become the laughing stock of the profession throughout the world, were it to open the door promiscuously to all these parasitic outgrowths of the healing art." Allopathic physicians went so far as to submit a false report to the board from supposedly well-recognized homeopaths that testified to the inadequacy of homeopathy. The report not only failed to sway the nonphysicians on the board; it also alienated them. Judge Joseph S. Bosworth, member of the board, dismissed it as a political attack, and board president Jackson Schultz regarded it as "a mere bundle of opinions" (quoted in Bowers, 1868, 411). When this failed, regulars drew on the AMA's no consultation clause to argue that inclusion of homeopaths would place those allopathic physicians on the board in violation of professional ethics. Beholden to the clause, allopathic participants would have to sever themselves from the profession or resign from the board. The board was, therefore, risking the loss of all allopathic cooperation.

Unfortunately for allopathy, homeopaths had considerable support on the board and in the local press. George W. Bradford, a regular physician on the board, lamented in his report to the Medical Society of the State of New York (1867b, 41–42):

> It was so found that a strong element in favor of Homeopathy existed in the Board from the start. The President of the Health Board, the President of the Metropolitan Police Commission, were strong adherents of Homeopathy; and another member of the latter body employed as a family attendant one regarded as a leading exponent of its doctrines and precepts. We successfully opposed this element in its attempt to gain for its friends a foothold in the Board by the appointment of homeopathic inspectors, despite the leanings of some of the newspapers to the heresy, and the zeal of three homeopathic reporters, ever ready to catch at any remark detrimental to the claims of the causes which they had espoused.

While Bradford overstated the extent of the association of the "homeopathic reporters" with the actual sect, it was true that local reporting on the controversy favored homeopaths, depicting the actions of allopathy as a crass power grab—precisely the type of political maneuvering and patronage that the board was established to combat. The *New York Times*' (February 20, 1867, 4) call for homeopathic inclusion employed a rationale that mirrored, almost word for word, the rationale given by state legislatures in the repeal of licensing laws:

> It seems to us—without pretending for a moment to judge of the merits of different schools—that simple justice argues in favor of this petition. It is not part of the case to discuss now the relative merits of any system

> of medical practice that appeals to the faith of the community. All may be good in their way; none are infallible; each can show a record most convincing in terms; each endeavors now and again to show the others to be foolish and weak. All that needs be said can be summed up in few words. A very large number of the wealthy and intelligent people of New-York and Brooklyn prefer the practice of Homeopathy, the system has passed through the ordeal of much sharp criticism, and in spite of ridicule and opposition has become established on a firm foundation. It has its schools, colleges, dispensaries and clients; it makes converts, and its successful treatment of cholera, of the disease more especially of women and children, has gained for it the confidence of large number so families.

Those on the board sympathetic to homeopathic inclusion, like President Schultz, were in a difficult position. On one hand, homeopaths made a compelling argument for inclusion that drew on the very arguments that justified the independent board's existence. On the other hand, the board could not risk alienating allopathic physicians to such a degree that they refused to cooperate. In the end, they compromised. Homeopaths would not be granted control of an entire cholera hospital but would be granted positions within the hospitals and assigned certain wards in which treatments were under their control. This compromise, unsurprisingly, left both protagonists unhappy. Homeopaths decried the fact that they would have to practice under the control of allopathy, and while they continued to petition the board for autonomous working conditions, they ultimately cooperated.

The board's recognition of homeopaths was more complicated for allopathic physicians, pitting rank-and-file regulars against the elite, sanitary-minded physicians. The no consultation clause, adopted by New York city and state medical societies under the guidance of the AMA, prohibited regulars from working with homeopaths and, in turn, raised questions about the fidelity of sanitary-minded physicians. NYAM went so far as to debate whether to withdraw its resolution of cooperation with the board. This debate centered on what to do about the allopathic members of the board who tacitly recognized the legitimacy of homeopathy by allowing this compromise. These members justified their acquiescence to the urgency surrounding the epidemic: "The sanitary committee, placed as they were in the position of soldier upon the eve of an important battle—to refuse to participate in the fray because another, whose principles they did not admire, insisted upon fighting at their side—would have been cowardice. The committee could not resign, because they were charged with a very grave responsibility as conservators of the public health" (Medical Society of the State of New York 1867b, 43). The rank and file pushed back, noting that "the interests of the public and the profession are too sacred to allow us to overlook for *this* reason an error of judgment which might have led to such disastrous results, and the tendency of which was to degrade the profession" (Hutchinson 1867, 68–69). Their argument reflected the AMA's general disdain of homeopathy—its system was absurd and therefore should not be indulged in any way, shape, or form. In the end, as in the case of the University of Michigan, the medical society avoided a major rift by narrowly defining the no consultation clause to only involve issues related to private practice. Board members, participating in government service, were not beholden to the no consultation clause. Rather than withdrawing its cooperation, the academy instead passed a resolution criticizing the board's position on homeopathy (Duffy 1968, 64).

Still the damage was done. The controversy over homeopathy solidified the rank and file's suspicion of public health. Immediately after the epidemic, allopathic physicians clamored to claim that the board of health had "originated in the Academy of Medicine and that the Health Bill should be regarded as the legitimate offspring of the Academy" (NYAM April 28, 1866, 7). But as the epidemic receded in memory and as the ecumenical colors of public health were revealed, allopaths distanced themselves from the board. It was no longer considered an allopathic offspring but rather a misguided endeavor susceptible to the pernicious influence of homeopaths. Even in New York, where reform efforts were initiated by a preeminent allopathic medical society, sentiments toward the board cooled.

CONCLUSION: A PLAN ABANDONED

In the end, the struggle between homeopathy and allopathy on the board followed a similar script to the debates over licensing, albeit with additional actors included in the fray. Regulars, initially viewing the boards as an opportunity to suppress homeopathy and promote their professional goals, saw these hopes dashed by a public institution unwilling to grant them a privileged epistemic recognition. Both the state legislatures and the boards of health supported a measure of transparency and inclusivity that clashed with the exclusionary program of the AMA. Just as democratic legislatures were suspicious of allopathy's claim that its knowledge should be granted a privileged position, so too were the boards. As subsequent chapters will show, the boards of health, which were held in high public esteem and possessed many resources, remained a key arena for the epistemic contest over cholera. However, the boards' understanding of disease as filth, their ecumenical organizational form, and their apolitical justification of epistemic authority prohibited regulars from harnessing them to advance their own professional agenda. They were too susceptible to democratic influences to provide allopathy's desired outcomes.

This episode in the epistemic contest makes clear that claims to epistemic authority can be framed along a variety of dimensions, and *how* they are framed restricts the possible professional claims that can be legitimately made in their name. Because the professions have become *the* way in which modern societies have organized expert knowledge (Abbott 1988), the relationship between epistemic authority and professionalization has been naturalized, so it is assumed to be not only unproblematic but also nearly synonymous. Justify one's knowledge, gain professional recognition. By showing how the manner in which epistemic authority was framed hindered, rather than advanced, the professional goals of allopathy, this case encourages researchers to decouple epistemic authority and professional strivings, recognizing them as distinct and separate (although often intertwined).

The fight over the boards of health (and, in turn, the definition of cholera as espoused by the boards) was waged on multiple fronts. Initially, it focused on entrenched urban politicians versus sanitarians. At this point, regular involvement was unofficial and restricted to elite sanitary-minded physicians. Sanitarians' claim to a privileged epistemic position rested on a disinterested, apolitical ethos toward knowledge. They successfully convinced state legislatures that public health should be removed from political calculus and placed under the control of those who could soberly assess the sanitary situation of the city and honestly carry out the requisite reforms. They won the authority to control cholera not on the content of their knowledge, nor on their particular epistemological stances toward medical knowledge, but on the basis of their *orientation* toward knowledge. Insofar as the new boards were seen as successful, allopaths benefited from their association with them. However, on two other fronts—allopathic physicians versus nonmedical actors and allopathic physicians versus homeopaths—the apolitical, disinterested justification stymied regulars' professional goals. Regulars were willing to adopt the rhetoric of disinterestedness when taking on political patronage, but less willing to extend this ethos in their interactions on the boards themselves. This made them vulnerable to attacks on the part of other sanitarians that their efforts to control the boards were political in themselves, contrary both to the intellectual ecumenism of public health and the ideals of democracy. The allopathic attempt to harness the boards for professional ends was framed by opponents as a bald political power grab, precisely the type of politics the boards were to be insulated from. This argument proved effective. While the legislatures agreed that the boards needed to be protected from political corruption, they did not want them to be insulated entirely from debate and democratic oversight. Instead, what the legislatures sought to do with the boards was to demarcate a politically untainted space in which debate could happen in a productive fashion. As such, regulars' exclusionary epistemology, and their attempts to impose this epistemology on others, fell on deaf ears.

Linking practices of exclusion to bad knowledge, homeopaths and plumbers effectively thwarted their powerful allopathic opponents. This chapter, therefore, demonstrates how intellectual disinterestedness as an ideal can be a resource for less powerful actors—a "weapon of the weak" (Scott 1985)—when advocating for inclusion within the context of democratic cultures. By drawing the analogy between democratic participation and intellectual ecumenism, plumbers and homeopaths effectively placed the onus on allopaths to justify their exclusion of others and their own privilege. The commitment to disinterestedness provided a platform for resistance by reframing issues of knowledge as issues of democracy, by situating them within a discursive space that politicized claims to intellectual privilege. In turn, the ethos of apolitical knowledge became, ironically, an effective political tool.

In the end, while the new boards of health offered enticing resources for allopathic professional goals, regulars were unable to harness them, leading many regulars to become skeptical toward public health. Undermined once again by the persistent tension between the exclusionary epistemology upon which allopathy asserted its professional claims and the democratic epistemology of government institutions, regulars had to ambivalently swallow their status as one voice among many.

4
CHOLERA BECOMES A MICROBE

THE CONVENTIONAL NARRATIVE OF THE "DISCOVERY" OF the cholera microbe reads as part medical mystery, part international brain race, and part microbial military campaign. In 1883, a smoldering cholera epidemic in Egypt raised the specter of yet another deadly pandemic. Barring some intervention, Europeans worried that the summer of 1883 would bring another season of cholera to their continent. This time, however, rather than await cholera's arrival, European governments decided to go after it. Armed with new tools of observation and a new germ theory of disease, they dispatched their best scientists, in the hope that cholera, perpetually elusive to quarantines, could be arrested by scientific acumen alone.

The Egyptian outbreak not only offered an opportunity to test new scientific ideas; it also held the potential for glory for burgeoning national scientific programs. European officials assembled elite scientific teams and sent them to Egypt to uncover the mysteries of cholera and win prestige for their country. France acted first, allocating fifty thousand francs for a cholera expedition. Dubbed *Mission Pasteur*, after the famous French bacteriologist Louis Pasteur, the French team included two of Pasteur's most promising assistants, Emile Roux and Louis Thuillier. The intrepid bacteriologists arrived in Egypt on August 15 and promptly occupied the best hospitals in Alexandria, where they carried out microscopic examinations of stool and blood specimens (Brock 1988). Their initial examinations led to sightings of a microbe that the expedition hoped might be the elusive cause of cholera. But the early leads went nowhere, and before anything could be confirmed, the cholera moved on from Egypt, having claimed somewhere between sixty thousand to ten thousand victims. One of these victims was the twenty-eight-year-old Thuillier. Reeling from the tragic blow of losing one of the brightest bacteriological minds in the world, the French expedition returned to Europe empty-handed.

Not to be outdone, Germany sent its own expedition on August 24, fast on the heels of the French team. Robert Koch, who had identified the tuberculosis bacillus the year before, was named the leader of the German Commission. Given great leeway to dictate the research program, Koch used the generous resources allotted by the German government to amass an impressive expeditionary force. Whereas the French team was outfitted only for microscopic and animal inoculation studies, the German team assembled "a complete travelling bacteriological laboratory" that included culture vessels, inoculation equipment, sterilization apparatuses, and other tools requisite for bacteriological examination (Brock 1988, 141–142). Koch was unambiguous about his goal; he wanted to validate bacteriology by isolating the organism that caused cholera. The extensive equipment amassed was necessary if the research was to fulfill Koch's stringent postulates for identifying the microbial origin of diseases.

Unlike its French counterparts, the German team succeeded. After the Egyptian epidemic abated, Koch followed cholera to India, departing on November 13. Within days of his arrival, Koch first observed the comma bacillus. After a few months of additional bacteriological and epidemiological research, Koch announced his discovery on February 4, 1884. Unable to reproduce the disease in animals, and thus failing to fulfill his own postulates, Koch knew that ironclad evidence of causation eluded him (Coleman 1987, 316). Nevertheless, he unabashedly reported, "It can now be taken as conclusive that the bacillus found in the intestine of cholera patients is indeed the cholera pathogen" (quoted in Brock 1988, 160). He went on to describe the bacillus as "not a straight rod, but rather . . . a little bent, resembling a comma. The bending can be so great that the little rods almost resemble half-circles. In pure culture these bent rods may even be S-shaped" (quoted in Brock 1988, 160).The innocuous appearance of this S-shaped microbe belied its deadly character, and it took a significant stretch of the nineteenth-century medical imagination to link the tiny organism to the destruction cholera wrought.

On May 2, 1884, Koch returned to Germany to a royal welcome. He met the kaiser and received a medal during an audience with Imperial Chancellor Otto von Bismarck. Koch had left for Egypt as a promising scientist; he returned a national hero. The bacillus, however, did not make the trip as Koch feared unintentionally introducing cholera to the Continent.

Was this the breakthrough discovery in medicine's tortured history with cholera? Did cholera become a microbe in 1884? Conventional histories answer both of these questions with a resounding yes (e.g., Chambers 1938; de Kruif 1996). According to these accounts, Koch's discovery was recognized immediately, and physicians worldwide mobilized in the name of the germ theory. News of it "spread to all of the laboratories of Europe and had crossed the ocean and inflamed the doctors of America. The vast exciting Battle of the Germ Theory was on!" (de Kruif 1996, 119).

In reality, the situation was far more complex. It would take time to sort out the meaning of Koch's research, time to configure it into the paradigmatic discovery of which future medical textbooks would write. At a basic level, it was hard for others to see what Koch saw, as his laboratory techniques had yet to be standardized. Only a handful of scientists in the world possessed the requisite technical skill to reproduce Koch's findings (Coleman 1987). Furthermore, bacteriology required a radical reorientation in the way most physicians thought about disease.

Whether cholera was a microbe thus depended on who was asked. Koch's research unleashed a firestorm of commentary. The British, sensitive to European criticism of their colonial empires and long averse to quarantines that inhibited free trade (Vernon 1990), scoffed at Koch's conclusion immediately and vehemently. The *British Medical Journal* dismissed the "bogey-germ" (quoted in Brock 1988, 150), while the English government published an official refutation (Ogawa 2000). Britain's beloved, pioneering nurse, Florence Nightingale, rejected the germ theory, as it seemed to undermine her efforts to improve sanitary conditions of hospitals (Rosenberg 1987a, 134). In France, Pasteur, perhaps out of competitiveness or wounded pride, raised questions as to the validity of Koch's claims. Even within Germany, Max von Pettenkofer, one of the world's most famous scientists, expressed skepticism.

The United States reacted to the news of Koch's research in conflicting and inconsistent ways. The *New York Times* (October 28, 1883, 8) seemed convinced, reporting, "The cable announced recently that Robert Koch, who went from Germany to study the Egyptian cholera epidemic, has found what many medical men of several nations have long looked for in vain—the cause of the disease." However, burying the article on page 8, the paper's editors hardly gave their vote of confidence to the newsworthiness of the event. In fact, Koch was only mentioned twice in the *New York Times* during the key years of 1883 and 1884. Coverage in other American papers displayed a similar pattern. Though reported, Koch's cholera research failed to provoke sustained headlines or to win a prominent place above the fold (Hansen 1999).

Still, the American press was more responsive than U.S. physicians, whose reviews were decidedly mixed. In the early 1880s, few members of the American medical profession were receptive toward "bacterian" ideas (Maulitz 1982).While Germany and France were well into a transition toward a university system based on specialization and modeled on scientific expertise, the U.S. educational system was still based on a more traditional theological model (Rudolph and Thelin 1991). Scientific research in the United States remained an amateur and community-based affair (Bender 1976), primarily carried out, not in universities, but in local learned societies (Oleson and Voss 1979). Far removed from European laboratory science, U.S. physicians were consumers rather than producers of medical research (Richmond 1954). Given this intellectual environment, American doctors were ill-equipped to interpret Koch's research and, in turn, its status as a discovery was by no means taken for granted. Bacteriological research produced in the laboratory demanded a shift in epistemic assumptions, and many American physicians were unwilling to make this leap yet. Instead, Koch's research got folded into the existing epistemic contest over medicine, where it was challenged and reinterpreted, in order to serve a variety of masters and ends.

THE MAKING OF A DISCOVERY

Contrary to the interpretations of triumphalist histories, Koch's discovery was neither self-evident nor unprecedented,[1] nor did it spread unimpeded throughout the developed world. Its status was more ambiguous, its dissemination complicated by fits and starts. In the United States, both homeopathic and allopathic physicians reacted to Koch's announcement in complex, often contradictory, ways. As with so many issues related to cholera, confusion reigned. Just because Koch spotted an S-shaped organism in his microscope did not mean American physicians would accept cholera as a germ. Much work still needed to be done to Koch's research by bacteriological reformers in order to make it palatable to Americans. Not only did U.S. doctors have to make sense of the strange new finding; they had to figure out what such research meant to the larger struggle over medical epistemology. The hero's welcome Koch would eventually receive in the United States in 1908 was a long way off.

The conventional histories of Koch's discovery offer little insight into the complex process by which the idea of cholera as a germ gained traction in the United States. In subscribing to a simple diffusion model, they treat Koch's research[2] as self-evident and self-interpreting. Koch's idea spoke for itself; the germ theory is portrayed as "an all purpose *deus ex machina*" that drove its own acceptance (Tomes and Warner 1997, 8). As discussed in the introduction, the diffusion model assumes a simplistic notion of scientific realism, whereby scientists uncover an objective reality, revealing the evident truth to the world. It misattributes agency to the "discovered" ideas, whitewashing away the ambiguity, uncertainty, and controversy often surrounding new ideas.

Even critical histories succumb to the dubious causality of the diffusion model. In his study of the social transformation of U.S. medicine, Starr (1982, 135) writes,

> The key scientific breakthroughs in bacteriology came in the 1860s and 1870s in the work of Pasteur and Koch. The 1880s saw the extension and diffusion of these discoveries, and by 1890 their impact began to be felt. The isolation of the organisms responsible for the major infectious diseases led public health officials to shift from the older, relatively inefficient measures against disease in general to more focused measures against specific diseases.

Here Starr falls prey to the crude causal explanations of the diffusion model. Uncritically accepting the bacteriological findings as "scientific breakthroughs," Starr describes their diffusion using the passive voice (e.g., the 1880s "saw the extension and diffusion of these discoveries"). The "isolation of organisms"—an idea—is given its own agency, as it "led" public health officials to certain measures. Letting ideas speak for themselves, these analyses offer limited insight into *how* these ideas came to be seen as paradigmatic discoveries. For even if we take for granted that a discovery is true (in whatever sense of the word), we still need to account for its acceptance. To avoid the tendency of ascribing to an idea "an ontological life of its own" outside of its historical emergence (Tomes and Warner 1997, 9), we need to historicize knowledge claims, embedding them in the context of their reception to unearth the processes by which actors advocate for ideas so as to get them institutionalized as discoveries.

Discoveries are not born. They are not unearthed in single moments of time but materialize over a long period *following* that moment. The sociology of scientific knowledge (SSK) has long criticized the folk understanding of a discovery "as a unitary event, one, which, like seeing something, happens to an individual at a specifiable time and place" (Kuhn 1962, 760). In practice, discovery is a social process involving two components—the production of a fact and the subsequent conferral of the status of discovery upon that fact. In relation to the first phase—the production of a fact—SSK, through laboratory studies, has produced a comprehensive body of research that illuminates the way in which scientists produce or construct scientific knowledge in practice (e.g., Knorr-Cetina 1999; Latour and Woolgar 1986; Pickering 1984;). It is the second phase—how the special status of "discovery" is conferred upon an idea—that remains underexamined. Discoveries do not spring up fully formed in the research process. Rather, they are defined as such through public struggles over the meaning of a given idea between actors with various agendas. In other words, discoveries are produced via work performed on an idea subsequent to its creation. The transformation of an idea into a discovery is a process that occurs literally after the fact.

Rather than conceptualizing discoveries as discrete events with agency of their own, this chapter examines the transformation of Koch's research into a discovery by embedding it within the epistemic contest over medical knowledge between regulars and homeopaths in the United States. One of the central problems with causal accounts that locate the efficacy of an idea in its content is the misguided assumption that the evaluation of ideas occurs within a universal epistemological system, in which facts are always facts according to some universal criteria. As I discussed in the introduction, this is a dubious assumption, bereft of any historical sensibility. Knowledge claims can only be judged—and only make sense—from within an epistemological system. Because medical epistemology was in flux throughout the nineteenth century, the actual process of reception of Koch in the United States is not a simple story of truth winning out; it is a story of struggle over basic epistemological assumptions. And discoveries can be a resource in an epistemic contest. The status of a discovery confers an importance and uniqueness upon an idea, something to be valued in and of itself. Because a discovery carries within it an implicit acceptance of certain epistemological assumptions, getting an idea accepted as a discovery can go a long way in capturing authority for one's epistemological system. In a sense, epistemological assumptions can ride the coattails of a widely respected discovery. In this case, Koch's discovery became a sort of Trojan horse that carried a commitment to the epistemology of the laboratory. Understanding how this idea was folded into a program of epistemological reform along the lines of the laboratory is necessary for explaining how Koch's research became a discovery in the United States.

How did Koch's research get reconfigured as a paradigmatic discovery for a new program of allopathic scientific medicine? Examining a case in which the production of an idea was relatively isolated from the context of reception allows for a targeted investigation of the interpretive work involved in the discovery process. Less concerned with what Koch did, I focus on the subsequent interpretations—and struggles over interpretations—of his research within the U.S. context in the decades after Koch's announcement. Adopting an attributional model of discovery (Brannigan 1981), this chapter demonstrates how Koch's research was transformed into an allopathic discovery that heralded a new era of scientific medicine controlled by regulars. Reformers within both medical sects staked claim to Koch, and both faced external and internal challenges in their attempts to align Koch's ideas with the preexisting systems of thought of their respective sects. In this "mnemonic battle" (Zerubavel 1999, 98), bacteriological advocates constructed discovery narratives that sought to situate Koch's research in the tradition of their respective sects, while simultaneously downplaying the ambiguity of Koch's initial findings through the production of promise. The first part of this chapter outlines the different narratives offered by homeopaths and regulars to show how the allopathic narrative of emergent discovery was more effective than the homeopathic narrative of prediscovery in providing a justifiable rationale for acting on Koch's finding.

The allopathic narrative provided an interpretive rationale for the embrace of bacteriology, but to solidify their ownership of Koch, allopathic physicians needed to supplement it with the organizational practice of building a network that linked them to Koch. The second section describes how allopathic reformers, building on their narrative justification, forged links with Koch and German laboratory science. In doing so, they claimed ownership of Koch's research and configured the idea—cholera as a germ—into a discovery that heralded a new era of medicine. For an influential subset of regulars, Koch's research became a discovery, and as owners of this discovery, allopathic reformers got to define the terms of the future of bacteriological medicine. When Koch became allopathic, homeopaths retreated into an oppositional stance that denied the legitimacy of the germ theory and that would prove professionally fatal.

An Attribution Model of Discoveries

Sociological theory on the notion of the discovery traces its roots to Robert Merton (1968), who noted the simultaneous emergence of the same discoveries arrived at independently by multiple scientists. Rather than singular, unique events, the existence of multiple discoveries suggests the influence of broader social and cultural processes on the progression of science. Additional research in the sociology of science further undermines the folk notion of discoveries by showing how scientific practice is inherently social (Shapin 1994; Shapin and Schaffer 1985), how laboratories actively intervene in natural processes, rather than passively observe them (Hacking 1983; Knorr-Cetina 1999), and how extrascientific factors influence scientific practices (Bloor 1991; Haraway 2006; Harding 1986, 1998). Discoveries are made, not unearthed.

However, the production of an idea is only the first step in the discovery process. To fully grasp how an idea becomes a discovery requires an analysis that moves "downstream" from the laboratory (Gieryn 1999). An attribution model of discoveries examines how discoveries are constituted over time via interpretive and organizational work done to ideas postproduction and, in turn, how the newly minted discovery is subsequently disseminated (Brannigan 1981). The emphasis is placed on the socially defined status conferred upon an idea—a status characterized by originality, singularity, and decisiveness. The "discovery-ness" of an idea is not inherent to the idea itself but is obtained through socially mediated interpretive practices. And the power of the status of a discovery is derived from its perceived significance. Contrasted to other possible statuses—replication or normal science—it transforms an idea into a watershed event that alters the future of knowledge.

There is a key temporal dimension to this status, as the event of the discovery demarcates the ignorant past from the promising future. To deem something a discovery is an exercise in marking time. The extraordinary present of the discovery reinterprets the past and anticipates a new future. The past becomes a period of ignorance and a repository of error, resolved and redeemed by the discovery. The future, on the other hand, becomes a rich new vista of possibility, emanating from the discovery. Because the attribution is a marking of a singular event heralding a new future, it confers upon its owner a certain degree of authority over this future. With this authority come professional, intellectual, and material payoffs to claiming a discovery for one's intellectual community. And as a vessel for particular epistemological assumptions, discoveries can be an important resource in epistemic contests. If one's idea is accepted as a self-evident discovery, it can naturalize the epistemological assumptions underlying it and, in a sense, smuggle them into legitimacy. With the future course of intellectual activity at stake, actors often jockey to claim discoveries to promote their epistemological vision.

As this chapter demonstrates, the configuration of a discovery involves both organizational and cultural practices. Actor Network Theory (ANT) has long recognized the importance of networks in the configuration and dissemination of discoveries, showing that the success of an idea depends on the creation of networks by which multiple actors are enrolled into the project of promoting it (Callon 1986; Latour 1987, 1988, 2005; Law 1992). In his exemplary study of the spread of Louis Pasteur's germ theory in France, Bruno Latour (1988) argues that Pasteur's ideas regarding microbes were successful because he was able to enroll allies, especially hygienists, in his project. This newly constructed network brought the laboratory into the field, disseminated Pasteur's germ theory, and promoted his research as a singular discovery. Network formation, as related to the ownership of a discovery, is particularly important in situations like the case discussed here when the production of the idea is severed from the context of its reception. For actors removed from the context of production, forming networks linked to those who produced it is essential in claiming ownership of an idea and using it to serve one's ends. In order for homeopathic or allopathic physicians to use Koch's idea in their epistemic contest, they had to form connections to Koch's laboratories, connections that were not predetermined, as both sects were similarly isolated from German science.

There is a crucial interpretative dimension required to establish the intellectual rationale for building networks. In other words, networks are necessary, but not sufficient, to configure a discovery. The transformation of Koch's research into a discovery first required significant interpretative work to be performed on the idea in order to make it acceptable—and accessible—to the very actors who would subsequently carry it along the network. Here is where Latour's account of Pasteurization stumbles. In a sense, he skips this intermediate interpretive step between the production of the idea and the construction of a network. Focusing primarily on practices of network-building, Latour largely ignores the practices of interpretation and sense-making necessary for building networks. In deploying military metaphors, he overemphasizes the building of alliances (Paul 1990), reducing actors' sense-making practices to the alignment of interests.[3] But ideas are adopted not only because they serve certain interests; they must also make sense to the actors adopting them. A discovery must be recognized, incorporated into, and reconciled with preexisting schemas. It also must find or prepare an audience willing to change its commonsense perceptions to accommodate it (Zerubavel 2003).

Because the attribution of a discovery marks a certain idea as a crucial event in time, the construction of narratives becomes a key interpretive practice for actors seeking to obtain the attribution of a discovery for an idea and to subsequently build a network around it. As analytical constructs, narratives make sense of disparate events by organizing them into a coherent relational whole through emplotment and attention to sequencing (Bruner 1991; Davis 2002; Polletta 2006; Riessman 1993; Tilly 2006; White 1980). They form a crucial component of the collective actors' repertoires in their struggles over meaning (Davis 2002; Patterson 2002). In narratives of discoveries, the "event" of the discovery is configured as the pivotal moment in the plot that demarcates the ignorant past from the promising future.

To transform Koch's research into a discovery, bacteriological advocates within allopathy and homeopathy constructed competing discovery narratives that sought to situate Koch's research in a manner that spoke simultaneously to the past and the future of their respective sects. Aimed internally at converting reluctant peers and externally at capturing ownership of Koch, these narratives emphasized certain elements of Koch's research that resonated with the particular traditions of allopathy and homeopathy and excluded those aspects that did not.

In constructing these narratives, allopaths and homeopaths had to walk a precariously fine line between embedding Koch within their respective traditions and arguing that the singularity of his idea demanded a departure from these very traditions. They had to balance a tension between continuity and discontinuity, framing the discovery as simultaneously *emanating from* the preexisting knowledge of their sect but *moving toward* a new knowledge base. To reconcile Koch with the past *and* anticipate a brighter future, the narratives deployed the logic of cognitive awakening (DeGloma 2010), in which the event of the "discovery" provided a flash of insight that demanded reformulation of the past and plotted a new course for the future. Through the construction of a new future, bacteriological advocates engaged in a type of *promissory practice* through which they glossed over the uncertainties and ambiguities in Koch's actual research to make bold claims (i.e., further discoveries, revolutions in medical therapies, and even the elimination of disease altogether) for its future implications. Ultimately, the allopathic discovery narrative proved more effective in resolving the continuity/discontinuity dilemma to successfully supply the rationale for an allopathic network linked to the German laboratory. Through the concatenation of the interpretive, narrative work of its advocates *and* the building of alliances with German science—this effective combination of cultural and organizational strategies—allopathic bacteriologists transformed Koch's cholera research into a discovery that promoted allopathy's interests in the epistemic contest.

NARRATING KOCH IN AMERICA

Early medical writings on the germ theory, Koch's research included, were plagued by inconsistencies in terminology, overzealous "discoveries" that later proved to be false, and technical difficulties in replicating findings (Hamlin 2009; Richmond 1954; Tomes 1997; Tomes and Warner 1997). In many ways, bacteriology raised more questions than it answered (Rothstein 1992, 267), as early bacteriological research contained a high degree of ambiguity. Ideas, and especially muddled ideas, cannot speak for themselves. Koch's identification of the cholera microbe was no exception to this rule of uncertainty, as it failed to meet his own rigorous standards. Koch established four postulates a microbe must meet in order to prove a causal link to disease:[4]

1. The microorganism must be found in abundance in all organisms suffering from the disease, but not in healthy organisms.
2. The microorganism must be isolated from a diseased organism and grown in a pure culture.
3. The cultured microorganism should cause disease when introduced into a healthy (animal) host.
4. The microorganism must be isolated again from the diseased experimental host and identified as being identical to the original specific causative agent.

Koch's cholera research did not meet the third postulate as he could not reproduce cholera in a healthy organism (Gradmann 2009). This failure was fodder for critics who argued that Koch's causal claims were exaggerated and overstepped the bounds of his evidence. Subsequent attempts to inoculate animals were plagued by technical issues and were viewed with wide skepticism (Coleman 1987; Ogawa 2000; Rothstein 1992).

To overcome this, Koch supported his microscopic research with epidemiological evidence, while drawing analogies to other diseases (e.g., leprosy and typhoid fever) widely believed to have a microbial cause that also failed to meet the third postulate (Porter 1998). Cobbling this evidence together, he radically simplified cholera, equating "all questions of cholera's cause with the doings of the bacillus" (Hamlin 2009, 215). His causal argument rested on a presence/absence reasoning, which argued that the bacillus was a necessary cause because in its absence, there was an absence of cholera, other things being equal (Coleman 1987). However, even this presence/absence argument was undermined by the presence of cholera bacilli in "healthy carriers," violating the first postulate.[5] The healthy-carrier problem would not be solved until the early 1900s with developments in immunology.

Koch's critics picked up on these problems. These critics were not marginal figures in the world of nineteenth-century medical science. Pettenkofer, perhaps the world's most famous German scientist (Richmond 1954), accepted Koch's bacillus as one important factor in the development of cholera, but he believed that the germ itself was harmless unless it underwent a process of fermentation caused by environmental factors in a type of poisonous metamorphosis. In 1892, to prove his fermentation theory, Pettenkofer publicly swallowed the comma bacillus to demonstrate its harmlessness, taking his failure to develop cholera as conclusive evidence of the microbe's inherent innocuousness. His fermentation theory gained widespread support in the United States (Richmond 1954). Louis Pasteur himself, a major proponent of the germ theory, remained skeptical of Koch's research (Brock 1988). And most British physicians, from whom American doctors had taken their cues for decades, opposed Koch's findings out of nationalistic pride and commercial concerns about trade.

There were good intellectual reasons to question the germ theory beyond the specifics of Koch's research, as it flew in the face of much medical common sense. For one, it was inconsistent with medicine's focus on predisposition; since medicine lacked any notion of the immune system, it seemed arbitrary to attribute the cause of disease to a small microorganism irrespective of the traits of the individual (Rosenberg 1987a, 138). It seemed to contradict a basic, widely accepted empirical observation—cholera affected some types of people (e.g., the poor, the malnourished, etc.) more than others. The germ theory seemed to also deny the relevance of the atmosphere and filth, the focus of public health for decades and the organizing principles behind most sanitary measures. And finally, skeptics had difficulty accepting the seeming randomness and meaninglessness of attributing the cause to a small microorganism (Rosenberg 1987a, 138). The skepticism toward the germ theory "was neither irrational nor reactionary; it was a reasonable position, taken by many leaders of the profession" (Rothstein 1992, 267).

In the United States, most of the opposition to Koch arose from a hesitancy to confer privileged epistemic status to laboratory analysis. During this period, Americans lagged far behind Europe in the medical sciences (Richmond 1954). The microscope and the laboratory were foreign to the overwhelming majority of American physicians, as most schools did not teach microscopy until the 1890s (Rothstein 1992). In fact, before 1884, there was no English-language medical text that discussed bacteriology in a comprehensive way (Bonner 1963). Given their unfamiliarity with the laboratory, American doctors were reluctant to grant primacy to it over other forms of evidence (Tomes and Warner 1997). Why, critics asked, should the bacteriological cause of cholera trump other known causes (e.g., poor sanitary conditions) gleaned from different types of data (e.g., sanitary surveys)? Drawing on clinical experience—in which doctors treated cholera patients at close proximity without succumbing to the disease—and epidemiological data—which pointed to environmental factors as causes of the disease—these critics argued that, while the bacillus might be correlated with cholera, there was no evidence to warrant Koch's causal claims: "There is no doubt, then, some relation between these bacilli and the cholera process. Nevertheless, the coincidence of the process with the bacilli does not prove the bacilli are the cause of cholera. The inverse may be true" (Kinsman 1886, 528). The comma bacilli might be thought of as "a concomitant effect rather than as the cause of cholera" (Blake 1894, 881). This "innocent bystander" argument, formerly used to reject bacteriological accounts of tuberculosis (Maulitz 1979), was common among Koch's critics.

For all these reasons, many American physicians adopted a wait-and-see policy toward the germ theory. An article on sanitation in *Harper's Magazine* explained their caution: "It is, indeed, claimed by some that the causation of certain diseases by specific organisms of microscopical minuteness has been demonstrated; by the majority of medical thinkers, however, the demonstrative evidence is not considered as complete" (quoted in Richmond 1954, 430). Preferring a middle course between complete adoption and outright rejection, most doctors suspended "judgment until more light had been thrown upon the subject by further investigation" (Fitz 1885a, 169). For these skeptics, no evidence, in and of itself, would suffice, as accepting the findings from the bacteriological laboratory involved a redefinition of what constituted evidence in medical knowledge. It demanded a leap of faith, and many physicians resisted jumping.

However, while doctors reluctantly debated the merits of Koch's research, the germ theory began to gain a measure of popular support before physicians even agreed that it was valid (Duffy 1993; McClary 1980; Tomes 1998). The media played an important role in building public support for bacteriology (Hansen 1999; McClary 1980; Ziporyn 1988). Media reports found fertile ground for the support of the germ theory among the public who had long balked at the noncontagious arguments of doctors; for them, cholera had always seemed contagious (Rosenberg 1987b). Koch had merely located the source of this contagion. As the germ theory gained public recognition in the mid-1880s, social reformers of many stripes began to promote a code of behavior based on the germ theory, and through their efforts, avoiding germs would eventually become a credo of modern living (Tomes 1998).

These assessments, both popular and professional, unfolded in the context of an epistemic contest over medicine that had reached a stalemate. Some reformers within both allopathy and homeopathy began to see the germ theory as a potential resource in this contest given its growing public support, even despite all of the intellectual difficulties it presented. While it is difficult to reconstruct the motivations of physicians who embraced bacteriology, it is clear that much of bacteriology's allure lay in its promise to bring an end to the stalemate. Decades of ugly fighting between sects had taken their toll, manifested in doctors' profound despondency about the state of their therapeutics, and to some, laboratory analysis offered an escape from this era of therapeutic nihilism (Warner 1991). As is often the case, there was a generational dimension to the adoption of the new ideas. Younger physicians hoped that by associating themselves with the latest science, they would advance their career prospects—carving out a niche in the competitive economic environment of nineteenth-century medicine—while also improving the public image of their sect.

When Koch announced his finding, bacteriology was neither homeopathic nor allopathic. It was something new altogether, its status unclear. If a sect could capture a popular idea, it might be able to ride it into wider acceptance. However, to transform the research into a discovery that promoted a particular professional agenda, homeopathic and allopathic reformers faced two tasks. They had to wrest ownership of the idea from their competitors and then transform the uncertain research into a paradigmatic discovery that heralded a new future in medicine. To achieve these ends, to make Koch's research speak to the past and the future, they created distinct narratives of discovery.

Homeopathy's Narrative of Prediscovery—In retrospect it may be difficult to understand the homeopathic claim to Koch's cholera finding, especially since bacteriology would later be folded into allopathy's professional program to defeat homeopathy. However, both homeopaths and regulars reacted with a mixture of curiosity and skepticism to Koch's claims (Rothstein 1992, 275). This similarity betrays a fact unacknowledged in most historical accounts of the emergence of the germ theory that situate it within the confines of allopathy—the new findings both resonated with *and* challenged the intellectual framework of *both* homeopathy and allopathy. Homeopathic reformers, like their allopathic opponents, saw the potential of Koch's research to tilt the epistemic contest in their favor.

Soon after Koch's announcement, homeopathic reformers staked a claim to Koch's work. This claim was rooted partially in an appeal to nationalism. Koch was German as was homeopathy's founder, Samuel Hahnemann. With this "genealogical tactic" (Zerubavel 2011, 82), homeopaths claimed Koch as a codescendant through his shared German ancestry with Hahnemann. The shared national heritage reflected a deeper, more subtle intellectual heritage underlying Hahnemann and Koch. Produced by the same context, their ideas were related. This argument was resonant in the context of nineteenth-century medicine, as nationalism was implicated in the acceptance and rejection of medical theories (Brock 1988; Ogawa 2000; Vernon 1990). *Where* ideas came from mattered a great deal, and the fact that homeopathy shared a national origin with Koch was no small matter.

This genealogical argument provided the backdrop for the centerpiece of the homeopathic claim to Koch—its prediscovery narrative. This narrative reframed Koch and his findings as confirmation of Hahnemann's initial genius (Rothstein 1992, 278). In this case, Koch was simply restating (maybe in a more sophisticated, systematic way) what Hahnemann purported long ago. The narrative proceeded as follows: in the early 1800s, Hahnemann dared to challenge conventional wisdom on cholera specifically and disease generally, but was persecuted for his beliefs. For decades, homeopaths stubbornly maintained Hahnemann's tradition only to suffer a similar persecution at the hands of allopathic physicians. Finally, Koch rediscovered Hahnemann's ideas and thus vindicated the entire tradition. The logic of this narrative was one of prediscovery (Merton 1968). Hahnemann's was the initial "prediscovery," which had somehow failed to convince; Koch's research was a rediscovery of Hahnemannian insight under a slightly different guise. The intent was to use Koch's public popularity to vindicate Hahnemann's genius and the homeopathic medical system.

Specifically regarding cholera, Koch's findings were portrayed as further confirmation of Hahnemann's original conception of the disease: "The germ theory of disease was anticipated by Hahnemann. As early as 1831 he alleged 'that cholera was propagated by organisms which were conveyed by the air,' and he advised the administration of camphor in material doses 'in order to destroy these pestiferous microbes' " (Wood 1899, 109). This view of cholera espoused by Hahnemann sat awkwardly with his understanding of the vital force, an awkwardness that was never fully resolved. Still, this early presage of the germ theory was useful to homeopathic bacteriological supporters, as it allowed them to situate Koch within the homeopathic tradition. Koch's relevance lay less in the novelty of his claims, but in the support these claims conferred upon old Hahnemannian ideas. The homeopathic narrative linked Hahnemann to Koch and other bacteriologists who should "be elected to membership of the American Institute of Homeopathy for the patient but brilliant, unconscious confirmation of the truth which Hahnemann promulgated" (Maddux 1892, 9).

With one eye to the past, the prediscovery narrative also looked to the future, linking therapeutic implications of the germ theory to homeopathy's law of similars. Recall that homeopaths viewed all disease as an imbalance of the vital force. The goal of homeopathic therapeutics was to bring the vital force into equilibrium, by prescribing mild, infinitesimal drug treatments that complemented the ameliorative capacities of the vital force (Coulter 1973). Taken to their logical therapeutic conclusions, both the germ theory and the law of similars arrived at a similar destination—vaccination.[6] Vaccination involved giving minute doses of the disease itself. The dilution of the disease for vaccination squared with the homeopathic commitment to infinitesimal dosages. Likewise, the rationale behind vaccination—preventing disease by building up a patient's resistance to it—squared with the homeopathic emphasis on the vital force. Vaccination was "*homeopathy pure and simple*" (McClelland 1908, 5). In this narrative, then, Hahnemann provided the foundational thought—the law of similars—upon which the future of bacteriology rested. Linking Koch to this founding moment, homeopaths situated the germ theory within the homeopathic tradition.

Despite the prediscovery narrative, Koch's cholera research was resisted by many homeopaths. The narrative attempts to incorporate Koch into homeopathic tradition were frustrated by an internal division within homeopathy that revolved around tension between the reforming "lows" and the older, more conservative "highs." Ostensibly a debate over the dilution of medicine, the fundamental issue at stake was the degree of fidelity homeopaths should display toward Hahnemann's original ideas. Younger, liberal homeopaths—"lows"—embraced ideas from outside homeopathy and argued for more individual freedom in determining doses (Coulter 1973). Without memory of the past battles with allopaths, they wanted more latitude in choosing their intellectual influences and were even open to cooperating with allopathy. The new science of bacteriology was part of lows' attempt to open homeopathy to new influences, and as such, they were responsible for the homeopathic prediscovery narrative. The "highs," on the other hand, rejected accommodation to new ideas, worrying that it would blunt homeopath's critical edge and distinct identity. Assuming a fundamentalist stance toward Hahnemann's texts, the highs rejected any attempt to incorporate new bacteriological ideas as illegitimate cavorting with allopathy. Veterans of past legal skirmishes with allopathy, they were unwilling to suspend their distrust. But they also pointed to the real intellectual problems in reconciling the germ theory with homeopathy. Because homeopathy considered that the ultimate cause of disease would always remain unknown, highs thought the new focus on etiology was misplaced. In 1886, the *Homeopathic Physician* proclaimed that "those who seek material causes of disease aided by the microscope will seek in vain" (quoted in Coulter 1973, 354). The idea that all disease could be reduced to a germ was anathema to this emphasis on the vital force. Therefore, while lows tried to convert some highs by linking Koch to Hahnemann, their success was limited. These internal divisions plagued homeopathy for the remainder of the nineteenth century, especially as many highs left the American Institute of Homeopathy to set up the more orthodox International Hahnemannian Association (IHA), which quickly denounced the germ theory. Consequently, the status of the germ theory remained ambiguous within homeopathy.

Part of this failure can be attributed to the nature of the prediscovery narrative itself, as it imbued Koch's research with uncertain status vis-à-vis future implications. If Koch's research was simply a rediscovery of Hahnemann, why should it even matter for homeopaths who had long been committed to these ideas? In what way was the future program it promised different, relevant, or even exciting? Couldn't the same promise be obtained by simply continuing the homeopathic tradition? Aside from capturing some of the public popularity associated with Koch, the prediscovery narrative suggested to many homeopaths a call to maintain the status quo. This minimized the singularity of Koch's research and diluted the urgency surrounding the new "discovery." As such, the prediscovery narrative failed to resolve the status of Koch's research within homeopathy and prevented its full transformation into a singular homeopathic discovery.

Allopathy's Narrative of Emergent Discovery—Prior to the bacteriological revolution, allopathy embraced a radical bedside empiricism, selectively translated from the Paris School. In their repudiation of the rationalist systems of the early nineteenth century, allopathy sought to purge medicine of all speculative hypotheses and theories by making individual sensory observation the foundation for medical knowledge. This legacy of suspicion toward theory presented a problem for allopathic bacteriological reformers. The germ *theory* proved to be a tough sell. This was especially true as it demanded privileging laboratory facts over clinical bedside observation. Allopathy had staked its epistemological claim in sensory observation, what they could see, touch, and hear, what they witnessed regarding disease in situ. Laboratory analysis removed disease from its natural environment and employed new technologies to expose disease in controlled environments. Adoption of the germ theory would require an epistemological shift that many regulars were resistant to take.

Advocates of the germ theory initially attempted to fold Koch's findings into allopathy's radical commitment to empiricism by appealing to sensory observation. They brought the cholera microbe to skeptics so that they could *see* it for themselves. In 1884, Dr. E. C. Wendt presented a sample of cholera bacilli prepared by Koch to the New York Academy of Medicine, inviting members of the society to observe with their own eyes what Koch saw (*Boston Medical and Surgical Journal* January 15, 1885). In 1885, F. S. Billings, surgeon general of the U.S. Army, offered a similar opportunity to those interested. But it was difficult for allopathic skeptics to actually see the microbe (Billings 1885, 476). They weren't trained to do so. For those steeped in the tradition of sensory observation, the way in which these new tools altered sensory experience was confusing. Spots in a culture were foreign, as the overwhelming majority of U.S. doctors lacked access to labs and microscopes (Duffy 1993; Richmond 1954; Tomes 1997, 1998). One bacteriological advocate lamented that the new techniques were "of so elaborate and difficult a character that the number of competent witnesses must necessarily be very small" (Hamilton 1884, 492). Translating them into useful knowledge was not only beyond the skill set of many regulars; it was beyond their imagination. These technical and conceptual difficulties were expressed somewhat comically when Albert L. Loomis, a well-regarded New York physician and future president of NYAM (who later converted to bacteriology), complained when he initially came into contact with the germ theory, "People say there are bacteria in the air, but I cannot see them" (quoted in Fleming 1987, 72).

To overcome these difficulties, allopathic bacteriological advocates had to reconcile Koch's research with the still-dominant radical empiricism. To do so, they constructed a narrative of discovery that de-emphasized the theoretical component of Koch's research. Koch's research was portrayed as an emergent discovery, driven not by a commitment to the germ theory but by fortuitous empirical findings. What he saw determined the theory, not the other way around. In this "narrative of emergent discovery," spontaneity was the key theme. Koch went abroad with only his microscope and an abiding motivation to uncover the truth of cholera. Once there, his careful work unearthed the microbial origins of the disease. Excluded from this narrative was any acknowledgment of Koch's theoretical commitments: "Never before has there been so little theorizing with so much work, and the facts which have finally been accepted are such as have come to light only after a long series of carefully made and oft-repeated experiments" (Loomis 1888b, 56). Koch, who set out to Egypt and India expressly to confirm his germ theory (Coleman 1987), was transformed into an empirical observer who, through patient observation, became the spokesperson for an illuminated factual truth. Koch himself encouraged this myth, publicly framing his "discovery" along a similar empiricist line by downplaying the influence of theory on his research (Gradmann 2009).

Therefore, in an attempt to make Koch acceptable to the bulk of their allopathic peers, bacteriological advocates transformed his research into an ideal exercise in empirical observation through this narrative of emergent discovery. To consolidate this interpretation, reformers tied the discovery to existing allopathic knowledge, knowledge that had been achieved by empirical means. Advocates argued that the new laboratory facts were complementary to bedside observation: "Recent microscopical and experimental researches in Egypt and Calcutta, made at the expense of the German government, by Dr. Robert Koch, one of the most successful detectives of disease-causing germs, seem to demonstrate, what *general observation of the disease* has already indicated" (*Sanitarian* 1884, 108–109, emphasis added). In accordance with the dictates of empiricism, the defense of Koch's discovery was waged through appeals to the "logic of potent facts" (*Medical and Surgical Reporter* 1892, 459). By 1894, William Welch (1920a, 27) was stating,

> The evidence now, however, for many infectious diseases is no longer aprioristic, but is based upon incontrovertible observations and experiments relating to the causative microorganisms. We owe to Pasteur as the pioneer and then to Koch and to their followers, the great mass of positive evidence, which has introduced a new era in the history of medicine.

In the hands of its bacteriological advocates, the germ theory had morphed into a mass of potent facts and observations that spoke for themselves.

This narrative did not go unchallenged within allopathy. Much of the internal debate hinged on whether Koch's research was an exercise in theorizing or whether it was driven by empirical observation. For older physicians especially, Koch's findings represented just another problematic theory in a long history of reductionist theories. Veterans of past epistemic battles remained skeptical, arguing, "This theory is a very seductive one, and although investigations in order to establish it have been very extensively made, and the statements of some observers are strongly in its favor, yet the facts do not warrant us in accepting it" (Loomis 1888a, 119). Like their homeopathic competitors, allopathic bacteriological reformers faced stiff internal resistance, reflecting the real conceptual mismatch between the prevailing intellectual foundation of their sect and the new laboratory science. To avoid a similar fate of fragmentation, they needed to do more work to link the future of American allopathy to the famous German scientist.

The Production of Promise

The germ theory's inconsistencies with each sect's tradition perhaps could be overlooked if advocates were able to construct an enticing enough future. As Karl Popper (1992, 16) notes, "Scientific discovery is impossible without faith in the ideas which are of a purely speculative kind." To construct such faith, however, bacteriological advocates had to deal with the present uncertainty and ambiguity of Koch's research. Turning their gaze to an imagined future, they engaged in promissory practices to overcome the present limitations of the research itself. In other words, proponents of the germ theory in the United States sought to explain away the uncertainty surrounding Koch's research via an appeal to an imagined future in which the germ theory would eradicate all infectious disease. In this way, Koch's discovery became folded into a larger program of bacteriological reform, a type of "promissory science" that justified current investment through ambitious promises of returns (Fortun 2008).

Advocates adopted two rhetorical moves to create a sense of promise around Koch's research. First, they assured skeptics that any current uncertainty in Koch's research would be eliminated by future research. For example, F. S. Billings (1885, 476), a bacteriological expert on the swine plague, argued, "Although the remaining proof, cultivation of the disease by means of the bacilli, has not yet been made, still in the present early state of our knowledge of the subject, we have no reason for declining to accept the bacillus as the cause of cholera." Billings points to future certainty to be achieved via the promise of further bacteriological research so as to downplay the present deficiencies. An extreme variation of this practice was an appeal to the present uncertainty as the very justification for investing in bacteriology: "The contributions of bacteriology to the medical art are all the more remarkable, because its methods and processes are still enveloped in much mystery—mystery which teaches us to expect much from the further developments of the new science as it gradually disperses the fogs which now envelop it" (Eliot 1896, 92).

Second, advocates promised a wide array of practical payoffs in the near future. In 1887, before the New York Academy of Medicine, Dr. E. O. Shakespeare (1887, 478) claimed:

> In the discovery of a reliable and an easily applicable distinctive means of diagnosis of Asiatic cholera at the very first moment of invasion of a

> locality or a country, I consider that Koch has conferred upon the human race an inestimable benefit for, with the proper application of this means of diagnosis, health authorities and Governments will now more than ever before have it in their power to prevent an invasion of cholera and cut short the spread of an epidemic. The value of this discovery, as I have said before, I consider incalculable.

The diagnostic promise was particularly robust in that it required no commitment to Koch's shakier causal claims; if cholera was *correlated* with Koch's bacillus, then the laboratory identification of the microbe could be an effective diagnostic tool regardless of issues of *causality*. Cholera, a disease with such violent symptoms that it was rarely misdiagnosed anyway, became a disease whose presence or absence could be determined with greater certainty by laboratory analysis. This would eliminate the debates over the presence of the disease that seemed to plague each epidemic: "The possibility of recognizing the first case of cholera in any given community, no matter how slight or how abnormal its symptoms, is capable of proving of immense value in permitting appropriate measures to be taken to prevent the spread of disease" (Fitz 1885b, 197). Of course, fulfilling this diagnostic promise would require the investment of public health laboratories, which de facto meant an investment in the bacteriological paradigm itself.

Additionally, advocates envisioned a new future for sanitation, a rational program focused on locating and destroying germs. The identification of microbes would be the means by which sanitary science would become truly scientific. While admitting its limited effect on therapeutics, T. Mitchell Prudden (1889, 94–95), future director of the Rockefeller Institute for Medical Research, optimistically argued that "knowing definitely, as we do now, what causes the disease, how and under what conditions it spreads, and what will destroy the germs, we are to-day in a condition, wherever sanitary and proper quarantine regulations are efficiently carried out, to largely prevent the access of the disease to our country, to stay the progress of an epidemic at its very outset, and to promptly allay the panic which the advent of a mysterious and deadly scourge is so prone to incite." In 1893, the editorial board of the *Medical and Surgical Reporter* (March 18, 1893, 421) claimed, "With our present knowledge of the comma bacillus, and how to destroy it, we can act with intelligence and effect in our efforts to ward off the coming pestilence." In actuality, the acceptance of bacteriology in public health had few early practical effects (Marcus 1979), and medical historians generally agree that the most effective sanitary reforms of the nineteenth century came under the older miasmatic theory of disease discussed in the previous chapter (Dubos 1987; McKeown 1976, 1979).

Beyond these diagnostic and preventative claims lay more tenuous predictions of prophylactic and therapeutic breakthroughs. Because Koch had identified the enemy, therapeutics could now be targeted toward combating the specific bacteriological culprit. Here was the dream of specificity, whereby specific treatments would attack specific microbes, the "magic bullets" later popularized by Paul Ehrlich, the German scientist who developed Salvarsan, a cure for syphilis. Advocates claimed that while it "cannot be claimed that it [Koch's discovery] has as yet aided the therapeutics of cholera," there are reports that "indicate that the problem of protective vaccination against cholera is nearly solved" (AMA 1892a, 529). These reports were only eluded to, not cited nor addressed directly. In 1892, the *Journal of the American Medical Association* (AMA 1892d, 757) predicted, "To-day bacteriology is about to revolutionize medicine by elaborating a specific treatment of the infectious diseases." Koch evoked visions in which a specific treatment would be developed for each specific germ. In the most expansive terms, Koch's discovery would result in nothing less than the eradication of the threat of cholera: "If, then, the germ theory of cholera be true, and it be a contagious and portable disease, then, as a matter of course, it must be admitted that it is a preventable disease" (Dixon 1885, 417). The discovery was reframed as so conclusive that "it will hereafter become actually a criminal act for any civilized government to permit of a spread of this highly infectious disease [cholera]" (AMA 1892c, 442).

Both homeopathic and allopathic advocates offered these promises in an attempt to obscure the ambiguity in Koch's findings and to foster a sense of anticipation. Both of their narratives attempted to construct a future for their sects based upon Koch. The initial appeal of the germ theory to its advocates "could be described in large part as almost an expression of faith; that is, the appeal was not based entirely on tangible evidence: to believe in the germ theory was to be optimistic" (Romano 1997, 54). In terms of efficacy, however, the logic of the allopathic narrative of emergent discovery reconciled epistemological tensions within the sect better and painted a more convincing future for reluctant physicians. While the logic of prediscovery could be read as a call to maintain the status quo for homeopathic skeptics, the allopathic narrative reframed Koch's research as unimpeachable empiricism. The basic premise of empiricism is that good discoveries should be pursued. Even if orthodox skeptics might reject Koch's specific findings, his approach and method were unassailable and thus worthy of further investigation. By tying a theoretically sanitized version of Koch's research to a brighter future, allopathic bacteriological advocates provided an interpretive schema that aligned Koch with allopathic interests in the epistemic contest. As such, the skeptics were willing to withhold final judgment and entertain the possibility that future facts would emerge to confirm Koch. It was through this faith in a future pregnant with scientific possibilities that regular advocates of the germ theory transformed Koch's research into a watershed discovery that opened a promising new horizon, one free of homeopathic challengers.

A NETWORK TO THE LABORATORY

Allopathic reformers solidified their narrative claim to Koch by forging an organizational network to German science. The establishment of such a network was built on the interpretive practices discussed above, as the allopathic narrative served as the justification for the network. In other words, creating a network only made sense if there was sufficient justification for seeing bacteriology as related to allopathic practice; the narrative supplied this rationale. Even if this justification was contested, it was a justification nonetheless.

Still, to consolidate control over Koch's research and its status as a discovery, young elite reformers within allopathy began a conscious program of building networks with German laboratory science. As the previous generation had traveled to Paris for advancement, a new cadre of elite doctors traveled to Germany to learn under Koch, creating an "invisible college" (Crane 1972) of bacteriological advocates who returned from Germany bent on reorganizing U.S. medicine along the lines of the laboratory sciences. Building on their narrative of emergent discovery, allopathic reformers constructed a network linked to Koch that would eventually transform the microbial account of cholera into a discovery owned by regulars.

The German-American network that would emerge at the turn of the century was not built from scratch in 1884. American doctors already had some links in place. Prior to the rise of bacteriology, some American doctors, consisting mainly of German-American physicians—sons of German immigrants making use of immigrant networks—traveled to Germany to take advantage of educational opportunities unavailable in the United

States. However, while the emigration of American doctors to Germany began as early as the 1850s, it was not until the emergence of bacteriological findings that Germany came to be seen as an alternative to Paris (Bonner 1963). The second wave of American doctors traveling to Germany was driven by a reforming impulse. In search of postgraduate training in German labs, approximately fifteen thousand elite doctors, mainly from the eastern United States, traveled to Germany to study medicine between 1870 and 1914 (Bonner 1963, 23). This group of doctors would transform the meaning of the German experience for U.S. doctors from one centered on educational opportunities into a radical program of reform for U.S. medical science through the laboratory. More than its sheer size, it was the composition of this latter group that would facilitate the rise of bacteriology in the United States. These were the future leaders of the profession. For example, virtually all Harvard medical school faculty members of the late nineteenth and early twentieth centuries spent time in German universities (Bonner 1963, 63). These German-trained doctors would eventually use their positions of influence to disseminate bacteriological thinking and to link their organizations to the German lab.

An analysis of the backgrounds of the most important American allopathic bacteriological reformers uncovers a persistent pattern of German connections, and more specifically, connections to Koch (Bonner 1963). William Welch, the stout bacteriologist and indefatigable medical reformer, is the exemplar. No individual played a more important role in reorganizing American medicine (Fleming 1987). An early supporter of laboratory analysis, Welch received training in Germany, though not in bacteriology. Upon returning to the United States, he ran a laboratory at Bellevue Hospital in New York. Although poorly funded (it lacked even a microscope), Welch's Bellevue lab was one of the few medical labs in the country at the time. In 1885, just a year after Koch's announcement of his discovery, Welch was offered the opportunity to establish a medical school at Johns Hopkins University. That summer, Welch left his lab in New York and traveled to Germany specifically to study bacteriology under Koch. He returned a committed bacteriologist and sought to mold the Johns Hopkins medical school along the lines of the German university. In doing so, Welch (1920b, 43) carved an institutional space within American medical education for laboratory research: "I believe it would do much to advance medical education and to encourage original research in medicine in this country, if the way were more freely open for academic careers in the sense in which it is in the German universities." Central to Welch's vision was the laboratory, as "well-equipped laboratories are essential to medical education" (Welch 1920b, 45). Johns Hopkins would become a key node in the network of U.S. bacteriologists, and Welch a leading figure in the bacteriological revolution in U.S. medicine.

Welch was a consummate institution builder. He was involved with not only Johns Hopkins but also the Rockefeller and Carnegie Foundations and several public health ventures. As such, he was indispensable in connecting U.S. medicine to the German program and in disseminating its vision of medicine through the training and placement of his students. Through Hopkins, he would socialize a generation of elite physicians into the bacteriological method and the laboratory ethos. His colleagues and students, the "Welch Rabbits," would dominate medicine at the turn of the century, occupying leadership positions in a number of important organizations, which they reformed along bacteriological lines (see table 4.1).

Some Welch rabbits, from the Wellcome Library, London.

Table 4.1. Welch's network*

Name	Relation to Welch	Position(s) held
Alexander C. Abbott	Student	Professor, University of Pennsylvania; Director of the Laboratory of Hygiene, University of Pennsylvania
Stanhope Bayne-Jones	Student	Professor, Johns Hopkins; Professor, Yale University School of Medicine; Dean, Yale University School of Medicine
Charles R. Bardeen	Student	Professor, Johns Hopkins; Professor, University of Wisconsin–Madison; first dean of the University of Wisconsin Medical School
Hermann M. Biggs	Research colleague	Director of Public Health in New York City
Joseph C. Bloodgood	Student	Surgical pathologist, Johns Hopkins
Charles V. Chapin	Student	Health Officer, Providence, Rhode Island (1884–1932); President, American Public Health Association; author, *The Sources and Modes of Infection* (1910)
Rufus Cole	Student	Director of the Hospital of the Rockefeller Institute for Medical Research
William Councilman	Research colleague	Professor, Harvard University; first president, American Association of Pathologist and Bacteriologists
Thomas S. Cullen	Student; Faculty colleague	Professor, Johns Hopkins
Abraham Flexner	Student	Education reformer; author, *Medical Education in the United States and Canada* (1910)
Simon Flexner	Student	Professor, University of Pennsylvania; first director of the Rockefeller Institute for Medical Research
Ernest Goodpasture	Student	Rockefeller Fellow; Professor, Harvard University; Director of William H. Singer Memorial Laboratories in Pittsburgh, Pennsylvania; Professor, Vanderbilt University School of Medicine; Director, Armed Forces Institute of Pathology
Christian A. Herter	Student	Professor, Bellevue Hospital Medical College; Professor, Johns Hopkins; cofounder, *Journal of Biological Chemistry*
William G. MacCallum	Student; Faculty colleague	Professor, Johns Hopkins
Franklin P. Mall	Student; Faculty colleague	Professor, University of Chicago; Professor, Johns Hopkins
Dorothy Reed Mendenhall		Pediatrician; first female graduate of Johns Hopkins

Eugene L. Opie	Student	Professor, Washington University School of Medicine; Dean, Washington University School of Medicine; Director, Phipps Institute for the Study and Treatment of Tuberculosis, University of Pennsylvania
William Osler	Faculty colleague	Physician-in-Chief, Johns Hopkins; author, *The Principles and Practice of Medicine* (1892)
T. Mitchell Prudden	Professional colleague	Professor, College of Physicians and Surgeons ; Professor, Columbia University; Scientific Director, Rockefeller Institute
Theobald Smith	Colleague	Researcher, U.S. Department of Agriculture; Professor, Columbia University; Professor, Harvard University; Director, Pathology Laboratory of the Massachusetts State Board of Health; Director of the Department of Animal Pathology, Rockefeller Institute
Walter Reed	Student	Professor, Army Medical School; head, U.S. Army Yellow Fever Commission
Francis Peyton Rous	Student	Instructor in Pathology, University of Michigan; Director of Laboratory for Cancer Research, Rockefeller Institute; Nobel Prize in Physiology or Medicine (1966)
George M. Sternberg	Professional colleague	U.S. Army Surgeon-General
Henry P. Walcott	Professional colleague	Chairman, Massachusetts State Board of Health; President, American Public Health Association (1886)
George H. Whipple	Student	Professor, Johns Hopkins; Professor and Director of the Hooper Foundation for Medical Research, University of California; Dean of Medical School, University of Rochester; Noble Prize in Physiology or Medicine (1934)
Whitridge Williams	Student; Faculty colleague	Professor and Obstetrician-in-Chief, Johns Hopkins
Milton C. Winternitz	Student	Chair of Pathology, Yale University; Dean, Yale School of Medicine
James Homer Wright	Student	Chief of Pathology, Massachusetts General Hospital

*This list is not intended to be exhaustive but rather illustrative of the extent of Welch's influence. His network was vast, and I have identified only those individuals who rose to positions of prominence. Additionally, not all positions each individual held are listed here.

Welch was linked closely to another major U.S. reformer and a future director of the Rockefeller Institute, T. Mitchell Prudden. A possibly apocryphal story holds that when studying with Koch, both Welch and Prudden were warned by Koch not to bring cholera cultures back to the United States for fear of unleashing an epidemic. Unable to contain their excitement, the young doctors each stole a cholera culture from Koch's lab. However, Welch and Prudden separately reconsidered the ramifications of transporting the bacillus to the United States, unintentionally running into each other on a Berlin bridge the night before their departure to dump their samples in a river (Fleming 1987, 76).[7] True or not, this single tale involves four of the major players—Koch, Welch, Prudden, and cholera—in the bacteriological reform of American medicine.

The network first introduced the laboratory into public health. In an effort to turn back the ambivalence many regulars had developed toward the intellectual ecumenism of public health, allopathic bacteriological reformers hoped to use the prestige of laboratory science (Rosenkrantz 1974). Those leading this effort were rooted firmly in the U.S./German laboratory network. Hermann Biggs, a student of Welch's at Bellevue who would later go on to head the Division of Pathology, Bacteriology and Disinfection of the New York City Health Department, studied bacteriology in Germany, picking up additional bacteriological bona fides by studying with Pasteur in France. Biggs and Prudden would become the first public health officials to put Koch's research to practical use, isolating the comma bacillus aboard a vessel in the port of New York in 1887. Through his actions in the health department, especially in combating diphtheria, Biggs became one of the most vocal advocates for bacteriological reform (Hammonds 1999). He remade the department along bacteriological lines, in an attempt to transform the laboratory into the obligatory passage point for all subsequent sanitary knowledge and interventions in New York. Other cities adopted this model (Jardine 1992).

Institutionalizing the laboratory in the United States would take great time and effort, but with this growing network, reformers had laid a foundation, culturally and organizationally, for a reformulation of allopathic medical epistemology. For these reformers, the new network meant more than a mere association with particular research or specific scientific figures; it highlighted their commitment to a new vision of medical knowledge centered on the laboratory. Under this new epistemology, the laboratory was not just another source of medical knowledge; it was to become the preeminent and privileged space for the development of that knowledge. Medical science would radiate outward from it. As such, the German program became synonymous with laboratory science, and laboratory science became synonymous with the future of medical science that the United States needed to embrace. As Welch (1920b, 74) argued:

> The supremacy of Germany in science is due above all else to its laboratories, and no more fruitful record of scientific discoveries within the same space of time can be found than that afforded by this laboratory during Koch's connection with it from 1880 to 1885. Thence issued in rapid succession the description of those technical procedures which constitute the foundation of practical bacteriology and have been the chief instruments of all subsequent discoveries in this field, the determination of correct principles and methods of disinfection, and the announcement of such epochal discoveries as the causative germs of tuberculosis—doubtless the greatest discovery in this domain—of typhoid fever, diphtheria, cholera, with careful study of their properties.

Although not perfect—"they have had the misfortune to produce a Hahnemann" (Squibb 1877, 6)—the Germans were unsurpassed in medical research. Allopaths thus traded Paris for Berlin. By linking themselves to Germany, U.S. reformers sought to put Germany's growing prestige in the service of the epistemic contest over U.S. medicine. The laboratory promised a brighter future for American medicine, for "it would be to the present epoch that posterity would look back as the time when those signal discoveries were made which led to the final adoption of the parasitic theory of the origin of all infectious diseases" (Flint 1884, 422). Through its narrative of discovery and the grounding of such narrative in an international network, reformers had transformed Koch's cholera research into one of these "signal discoveries."

CONCLUSION: POLARIZATION AND THE LAB

Contrary to the heroic narratives, the 1884 "discovery" of the comma bacillus was not *the* defining moment in resolving medical debates over the disease and deciding the allocation of epistemic authority in medicine. American germ theory advocates still had much work to do in consolidating the authority of the lab and convincing others of the veracity of bacteriology. However, by 1894, a decade after Koch's announcement, reformers had produced an account of Koch's research that configured it as a discovery and located it solidly within the allopathic tradition. Furthermore, they had created a network infrastructure in which the status of this discovery was taken for granted. For at least one important subset of allopathy, Koch was king and cholera was a microbe. In this network, infectious diseases had been transformed into entities that could only be identified in a laboratory through the determination of the presence or absence of a specific microorganism (Cunningham 2002).

Full medical reform along the lines of the germ theory and the laboratory would require further time, investment, and resources. Given allopathy's checkered history with government institutions, the procurement of such resources remained a question. Nevertheless, while widespread consensus around bacteriology would not be achieved until the 1910s, the groundwork for this revolution had been laid. Allopathy defined and established the terrain for future medical and professional debates, having wrested Koch's findings from homeopaths. By establishing networks to Koch's laboratory and promoting bacteriology, reformers began the transformation of Koch's research on cholera into a paradigmatic discovery for allopaths. If the germ theory of cholera was to gain recognition, it would be regulars who benefited.

To apply the attribution of a discovery to an idea is to eliminate its uncertainty by naturalizing it as a self-evident truth. Indeed, this chapter reveals practices by which bacteriological advocates attempted to eliminate uncertainty and ambiguity surrounding Koch's research. In many ways, they succeeded in rendering evaluations of it more definite. So what happened to the uncertainty of Koch's findings so evident in the 1880s? The conventional narrative would have us believe that it was eliminated by additional research. This claim fails on three accounts. First, much of the uncertainty surrounding Koch's findings centered on his inability to meet the third postulate—reproducing the disease in animal hosts. Subsequent attempts to achieve this result faltered and were met with skepticism. Second, advocates had to contend with the problem of healthy carriers; not all individuals infected with the comma bacillus succumbed to the disease. Opponents of the germ theory, like Pettenkofer, went so far as to purposefully consume the bacillus to deny its causal connection to cholera. Lacking an understanding of the immune system, researchers would not solve the problem of the healthy carrier until the twentieth century, when immunology developed. Finally, the heroic narrative fails to acknowledge the magnitude of the epistemological shift the germ theory required. Any account that depends on "evidence" as a causal explanation is limited, because it fails to acknowledge that what counted as evidence was contentious. Bacteriological advocates could not simply present their findings for evaluation; they needed to convince their peers that the epistemological assumptions underlying these findings were themselves legitimate.

The flaws in the conventional accounts demand we look elsewhere to explain the resolution of the uncertainty in Koch's findings, namely, to the interpretive practices that advocates deployed to downplay or obscure this uncertainty. The narrative and network constructed by allopathic bacteriologists created an invisible college, in which the germ theory of disease went unquestioned. From this intellectual and organizational foundation, a small group of influential, committed elites embarked on a program to convince others of the "discovery-ness" of Koch's research. Though not yet dominant, there was now a space within allopathy where Koch's findings were part of the taken-for-granted common sense. This was not just an abstract intellectual space; it was rooted in an organizational infrastructure of a handful of elite medical schools and boards of health. As advocates structured (or restructured) these organizations along the lines of the laboratory, the germ theory became integrated into the operating assumptions of these institutions. In the process, the theory's uncertainty was replaced by a certain unquestioning acceptance.

Uncertainty and ambiguity were also absorbed through a process of polarization that followed the establishment of the allopathic network. Regular reformers transformed Koch's research into an *allopathic* discovery. They folded Koch's particular finding into their professional project of medical reform. Koch became an allopath, and his findings became entangled in the epistemic contest between regulars and homeopaths in the United States. This newly acquired association resulted in the extreme polarization between regulars and homeopaths in their interpretation of Koch's findings. As the new owners of the idea, allopathic physicians tempered their criticism of Koch's claims. Newly contextualized within the larger body of bacteriological work, Koch's cholera research was not to be judged in isolation but rather as part of a system of bacteriological findings. Cholera was connected to other bacteriological "discoveries" (e.g., Koch's discovery of the tuberculosis bacillus, Pasteur's research on anthrax, the discovery of the diphtheria bacillus antitoxin in 1883). While each of these findings had its own problems, individual uncertainties were diluted in a larger collective of claims, as "discoveries" drew strength through mutual reinforcement and analogical reasoning. Advocates dealt with the deficiencies in the research on any one disease by arguing that limitations had been overcome in research on another disease. Uncertainty was reframed as a temporary situation. Moreover, as bacteriology became linked to an allopathic future that promised professional recognition, the stakes for critics within allopathy who questioned it increased dramatically. They could reject the germ theory, but now that the germ theory was solidified as an *allopathic* theory, they ran the risk of being associated with homeopaths or other quacks. The more professional goals were tethered to bacteriology, the less intellectual space remained within allopathy from which to criticize it.

For homeopaths, the ambiguity surrounding Koch's research was resolved in the opposite direction. Philosopher Hilary Putnam describes a phenomenon of intellectual competition called recoil. When an idea gets attributed to one party, the opposing party, "dominated by the feeling that one must put as much distance between oneself and a particular philosophical stance as possible" (Putnam 1999, 4), "recoils" from it completely, denying that the idea possesses any insight whatsoever. In essence, this is what happened for homeopaths and bacteriology. Now that Koch was associated with allopathy, there could be no more entertainment of his ideas. Koch was unambiguously wrong. The initial similarity in reactions of homeopathic and allopathic reformers gave way to a highly polarized situation in which the sects opposed each other. What was once seen compatible with homeopathy (and even synonymous with it) was now viewed as "diametrically opposed" to the teachings of Hahnemann (Tooker 1885, 6).

As the associational hue of the newly minted discovery rendered it unacceptable to homeopaths, many came to repudiate the legitimacy of the germ theory entirely, denying that Koch had discovered anything. But as public support for the germ theory grew, this position of denial became increasingly untenable and detrimental to homeopath's professional project. Shifting from a strategy of denial to containment, homeopaths sought to downplay the relevance of Koch's work, deeming it immaterial to therapeutic practice. They justifiably pointed to the germ theory's lack of therapeutic breakthroughs, which, at the end of the day, were the measure by which medical theories were judged. When it came to actually treating people, the germ theory was next to useless, nothing but "a delusion and a snare" (Tooker 1885, 20). In adopting this tack, homeopaths altered their own claims to medical expertise. In order to carve a niche for homeopathy in an environment increasingly distracted by the promises of the germ theory, they repositioned themselves in a more restrictive sense as experts in medical therapeutics.

Regardless of the strategy adopted, as bacteriology gained popular support and as allopathic reformers increasingly reorganized medical institutions along the lines of the lab, homeopaths found themselves in a precarious situation, having lost out to regulars in the battle over ownership of Koch. They had ceded the germ theory and its discoveries to allopathy and therefore excluded themselves from any claim to its future accomplishments. As disparate bacteriological findings coalesced into a new medical paradigm rooted in the laboratory, homeopaths lost prestige. Locked out of the lab, homeopaths became more and more marginalized from mainstream medicine. Having opted out of this process, they were relegated to the sidelines for much of the remainder of the epistemic contest over medical knowledge in the nineteenth century, passive spectators to the main events.

Once regulars had captured the discovery, assessments of it no longer centered on the content of the idea, but rather on who was associated with the discovery. To argue that this was a predetermined, natural outcome of regulars' preexisting intellectual commitments, as many conventional accounts do, is to engage in an impoverished exercise of reading history backward. The fact is that bacteriology, if it was to be adopted, was going to require some fundamental revisions of both sects' understanding of disease and a reformulation of the epistemological foundation underlying their disease models. The germ theory contradicted homeopathy no more than it contradicted the dominant understandings of disease among allopaths. The epistemological leap from radical bedside empirical observation to the interventionist laboratory was no greater (and probably less) than the leap homeopaths would have had to make.

In the end, the ownership of Koch was not inherent in his ideas. Koch's "discovery" was not created in India but rather downstream in the way in which American advocates struggled to claim it, to transform it into a discovery, and to link the discovery via networks to a more general system of intellectual reform. Allopathic reformers still had work to do to realize the future promise that they claimed the laboratory held. This future would be expensive, and reformers still had to figure out how to raise the requisite funds for their reforms. But once completed, after the laboratory was institutionalized and regulars had achieved epistemic closure through bacteriology, Koch's research would finally become what it is now remembered as—a textbook discovery.

CAPTURING CHOLERA, AND EPISTEMIC AUTHORITY, IN THE LABORATORY

By the final U.S. cholera epidemic in 1892, bacteriological reformers had gained an important foothold in municipal and state boards of health. The allure of the boards for allopaths, previously spoiled by their intellectual ecumenism, was reinvigorated by Koch's discovery. Many bacteriological reformers viewed municipal laboratories as *the* way in which regulars would finally capture control of the boards, defeat cholera, secure the exclusive right to define it, and ultimately wrest epistemic authority from homeopaths and sanitarians. Even though the boards had long been under the control of a diverse group of sanitarians, allopathic physicians had maintained a healthy presence on them and were able to divert some resources to establish municipal laboratories. Thus, the stakes for the 1892 epidemic were high, as it would test these municipal laboratories, which, if successful, would validate bacteriology and, in turn, allopathy.

The early efforts to integrate laboratory analysis into public health were part of a new strategy to combat cholera, and other infectious diseases—a strategy that boasted some government support. In the wake of Koch's discovery, the U.S. government wanted to clarify the "conflict of opinion" (Shakespeare 1890, 2) over the germ theory between Koch and the British government, which had vocally denounced his findings. By executive order President Grover Cleveland sent Edward O. Shakespeare to evaluate bacteriology in action during a European outbreak of cholera. In the autumn of 1885, Shakespeare, a bacteriologist himself, assembled a "traveling laboratory" and arrived in Europe. He met with Koch, along with other European bacteriological luminaries, and traveled to Spain to evaluate Dr. Jaime Ferran's anticholera inoculations. In 1890, he published his thousand-page *Report on Cholera in Europe and India*, a tome that served as a clarion call for a bacteriological approach to cholera in the United States. Arguing that "nearly every studious physician of experience" (Shakespeare 1890, 447) believed that cholera was a microbe, Shakespeare laid out a blueprint for a bacteriological attack on the disease that included a national system of maritime quarantine and interventions based on laboratory science. Public health needed to move beyond broad sanitation efforts to a focused program that targeted specific germs, for it was "folly" (Shakespeare 1890, 854) to continue to approach diseases as if they were noninfectious.

Shakespeare's blueprint focused on three potential benefits of the laboratory. First, the bedrock of the new program was the promise the germ theory held for diagnosis. Diagnosing contagious diseases in an era before laboratory culture methods was fraught with difficulty, as physicians had to "rely on empiric observations and broad experience" (Markel 1997, 41). It was an exercise in intuitive clinical judgment, and predictably, conflicts often arose that stalled interventions. While laboratory methods were not foolproof and only a few physicians were skilled enough to perform them, Shakespeare believed that they promised immediate and accurate knowledge of the disease's arrival. This *perception* of their certainty allowed reformers to claim that "the only absolutely positive means of diagnosis of Asiatic cholera" (Welch 1893, 4) was to "see" the microbe. Second, Shakespeare lauded the prophylactic potential of laboratory analysis, albeit with reserve (even though Ferran's data on his inoculations showed inconsistencies, Shakespeare remained bullish about its prospects). Finally, municipal laboratories could perform an invaluable educational service. If a lab were established "in every city of any size," all physicians would be exposed to the new sciences and become "acquainted with the results of modern laboratory investigation" (Chapin 1934a, 90–91).

The 1892 cholera epidemic presented an opportunity to put Shakespeare's vision to work. Once again New York was at the center of the drama, as its bustling port was the place most threatened by the disease. The infrastructure was more or less in place, as the city had experienced a cholera scare in 1887. When cholera failed to materialize that year, Hermann Biggs, the resident bacteriologist of the New York Department of Health, claimed it as a victory for bacteriology (Fee and Hammonds 1995) and parlayed it into ten thousand dollars' worth of municipal funding to create the world's first municipal diagnostic bacteriological laboratory. This laboratory became the center of New York City's public health efforts and the place through which all actions pertaining to cholera had to pass.

As cholera approached in 1892, the focus was on establishing an effective quarantine system based on laboratory science. The preferred preventive measure in the early 1800s, quarantine had fallen into disfavor during the nineteenth century, as cholera continuously evaded its grasp and public health officials, operating under the idea that cholera was filth, concentrated on basic sanitation. Bacteriology redirected attention back toward quarantine. According to Shakespeare's vision, bacteriological methods would prevent the cholera microbe from gaining entry into the city. A ship believed to be infected would be held up in the port. The lab would then test passengers to see if the comma bacillus was present. If the ship received a clean bill of health, it could dock. If not, the passengers would be removed to quarantine stations where they would be detained until the incubation period of the microbe was over and cholera could no longer be detected. Under this system, bacteriologists were poised to determine quarantine policies, as they were the only ones who could definitively identify cholera aboard ships and understood the life cycle of the bacteria that determined the length of the quarantine. Such was the logic of the new bacteriological regime.

Thus, the board of health, armed with its new laboratory, sought to contain cholera in quarantine and, in turn, tame the disease once and for all. However, instead of a crowning success, the 1892 epidemic became an unmitigated public relations disaster for city officials, the New York Department of Health, and by association, bacteriologists. Once again, bacteriologists came up against the very problem allopaths had faced in public institutions throughout the course of the epistemic contest—outside interference. During the epidemic, bacteriological recommendations regarding the quarantine were routinely ignored and rejected (Markel 1997). Rather than serving as a rational application of bacteriological principles to sanitary science, by which "the exact replaced the conjectural in this branch of medicine" (Osler in Thayer 1969, 128), quarantine policies were ineffectively jerry-rigged out of xenophobia and political squabbling between local, state, and national authorities (Markel 1997). The results were embarrassing, and quite publicly so. In the end, despite the promise of bacteriology, what reformers got was a political farce that forced them to rethink their professionalization strategy, to look for a way to avoid the state altogether in order to achieve epistemic authority.

On August 30, 1892, cholera arrived in port aboard the SS *Moravia*, a passenger ship from Hamburg. The *Moravia* had lost twenty-two passengers to the disease, whose presence was confirmed by laboratory analyses. Unconcerned, bacteriologists expressed confidence that the disease would not jump the quarantine (e.g., "With our present knowledge of the comma bacillus, and how to destroy it, we can act with intelligence and effect in our efforts to ward off the coming pestilence" [*Medical and Surgical Reporter* 1893, 421]), provided that protocol was heeded. It wasn't. Rather than identifying and separating infected passengers according to bacteriological principles, Port Officer William T. Jenkins chose to keep all passengers aboard the ship for two weeks, unnecessarily exposing healthy passengers to cholera. Unsatisfied with Jenkins's handling of the quarantine, President William Harrison jumped into the fray, declaring a twenty-day quarantine of all ships arriving in American ports. The length of the federal quarantine was determined by an economic, not bacteriological, rationale, as twenty days was a period long enough to make transport too expensive for shipping companies to bear, effectively putting a halt to all trade (Markel 1997, 98). Despite the fact that the economic logic behind the quarantine length contradicted bacteriological science, it was fortunately wrong in the right direction, as it was actually longer than bacteriologists recommended based on the incubation period of cholera.

The real problems arose when Jenkins refused to comply with the federal mandate. Despite seeking the advice of "four of the most eminent bacteriologists in the nation" (Biggs, Welch, T. Mitchell Prudden, and George Sternberg), Jenkins ignored their advice to coordinate his activities with the federal government, balking at what he perceived as undue federal meddling. Jenkins's rejection of federal oversight took a darkly comedic turn with the unfortunate handling of another cholera ship, the SS *City of Berlin*. Unlike the *Moravia*, this ship from Antwerp was determined to be free of cholera. Defying President Harrison's twenty-day quarantine, Jenkins gave the ship permission to dock. The U.S. Collector of the Port of New York, however, refused to grant entry to the *City of Berlin*. What followed was a "dramatic illustration of the confrontation between state and federal rights culminating in an almost comic sending of the unfortunate steamship up and down the bay between the quarantine station and the port collector, as each official refused to recognize the authority of the other" (Markel 1997, 99). Passengers were needlessly confined to the ship, as it yo-yoed up and down the port. Ignored during this entire squabble were the negative laboratory findings. Federal and local officials not only refused to recognize each other's authority; they failed to recognize the authority of the lab.

The problematic handling of the SS *Moravia* and the SS *City of Berlin* was only a prelude to the farce involving another cholera-infected ship, the *Nor-*

mannia, on September 3, 1892. Because the ship's passenger manifest contained a number of luminaries, including the famous British entertainer Lottie Collins, U.S. senator John McPherson of New Jersey, and newspaper editor E. L. Godkin, the press took a special interest in it, scrutinizing the intensifying conflict between Jenkins and federal officials. This created a circus-like media atmosphere. Reporters, not allowed to board the infected ship, commandeered small boats and conducted interviews with passengers by screaming across the water (Markel 1997). Godkin himself wrote a number of widely circulated letters that documented and condemned conditions on the ship.

Jenkins decided to keep the *Normannia* passengers aboard until a quarantine hospital could be procured. Once again healthy passengers were unnecessarily exposed to cholera. Furthermore, ignoring bacteriological advice, the port failed to provide clean water to the passengers, enabling the spread of the disease. Describing the scene on the ship, *Harper's Weekly* reported,

> In the mean time the *Normannia*'s 600 passengers were penned up in the ship with death lurking on every side of them; no water, save that from the polluted Elbe, until the second or third day of their detention; and almost every hour bringing from the steerage its story of a "new case." And those steerage passengers, unclean cholera-trafficking wretches that they are—heavens! what must have been the torture of the uninfected—huddled together in quarters reeking in filth, where sleep was impossible and the moments agonizing in expectation of the plague? (Whitney 1892, 920)

The magazine damned the quarantine as "utterly unnecessary and barbarous to a degree bordering on the fiendish" (Whitney 1892, 919).The *New York Times* (September 9, 1892, 1) described the sentiments of the passengers: "It is a game of life and death to them, and they cannot understand why the authorities should be so tardy in removing them from what they regard as an imminent source of danger—enforced existence where they now are." When the city finally designated a hotel on Fire Island as a site for a cholera hospital, the decision immediately set off explosive local resistance by the island's inhabitants. Local fishermen formed a mob, dubbed the "Clam Diggers" by the press, and threatened to burn down the hotel/cholera hospital. The Clam Diggers refused to let the ship dock, threatening riots and violence should cholera be brought ashore. Governor Roswell Flower had to call in the National Guard and the Naval Reserve to force the mob to back down. Even after officials secured their cholera hospital, they failed to heed the advice of bacteriologists. The hospital might have done more harm than good, as sanitary conditions in the hotel were filthy, especially the water supply. Once again, healthy passengers were exposed to an environment that only cholera could love—an environment not so different from that aboard the *Normannia*.

Throughout the 1892 cholera scare, bacteriological knowledge was continuously ignored, resulting in ineffective policies that were universally repudiated by the local and national press. Officials inconsistently applied quarantine and sanitary measures to different groups throughout the ordeal. Cabin-class passengers were given preferential treatment in deference to their class status; steerage passengers were not afforded the same level of concern (Markel 1997). Moreover, the debates over cholera took on an increasingly racial tone.[1] The link between race and disease was joined with nativist fears to produce a lethal public backlash against the Russian Jewish immigrants who were blamed for bringing cholera to the States (Markel 1997). Despite repeated proclamations from bacteriologists that the disease was not in any way related to race or nationality, sanitary efforts targeted Russian Jewish immigrants, deemed the "scum of invalided Europe" (Whitney 1892, 920). Once tied to place, cholera now was written over with ethnicity (Markel 1997).[2] In fact, the primary effects of the 1892 epidemic were felt, not in public health, but in immigration reforms; fear of disease provided justification for the restrictive anti-immigration policies of the early twentieth century (Markel and Stern 2002).

Although only 130 people died during the 1892 epidemic, the press universally denounced the official handling of it. There was to be none of the celebratory praise like that which the board received in 1866. What promised to be a crowning moment for the laboratory became an embarrassment, complete with bumbling officials, petty politicians, and victimized celebrities.

Bacteriological reformers drew two lessons from the ordeal. First, as was the case in 1866, the municipal boards of health were too susceptible to political manipulation—too open to outside meddling—to offer unambiguous professional benefits for allopaths. Second, if bacteriologists wanted to win the epistemic contest, they were going to have to search for a way to achieve it outside of state institutions, to create organizational spaces insulated from the vagaries of politics.

CIRCUMVENTING THE STATE

The inability to translate the 1892 epidemic into a victory for bacteriology, and thereby usher in a new era of medical science, highlighted the persistent professional obstacles facing allopathic reformers in public institutions throughout the epistemic contest. Public institutions like state legislatures and boards of health had long recognized input from a plurality of voices when it came to cholera. They refused to grant sole, or even final, authority to allopaths on the matter. The state's celebration of openness undermined allopathic legislative efforts for privileged recognition, resulting in a dismal record that amounted to "little other than a long panorama of heroic endeavor and humiliating defeat" (Markham 1888, 5). As early as 1870, the AMA was questioning whether its legislative actions were not entirely misguided: "Legislative enactments in the various States of this Union clearly show that no reliance can be placed on either the uniformity or permanency of any laws now relating to the practice of medicine" (quoted in Medical Society of the State of New York 1870, 39). The political corruption of bacteriological interventions during the 1892 cholera epidemic intensified calls for a new strategy and revealed that bacteriological reformers had much work to do if they were going to achieve a system of medicine under the control of the laboratory.

The external obstacles to their program were compounded by the reluctance of many rank-and-file regulars to embrace bacteriology. Between 1884 and 1892, a widespread consensus around the idea that cholera was a germ had taken hold within allopathy. The preeminent textbook of the period defined cholera as a "specific, infectious disease, caused by the comma bacillus of Koch, and characterized clinically by violent purging and rapid collapse" (Osler 1895, 132). But as the last chapter showed, the manner in which reformers had to alter Koch to make him palatable for most allopathic physicians downplayed the revolutionary demands of the germ theory. In smuggling Koch into the predominant radical empiricism as such, questions remained as to the significance and reach of bacteriology. Beyond diagnostic testing, what relevance did it have for medical practice? Furthermore, it was a long intellectual distance between the recognition of cholera as a germ to the view that medical knowledge should be radically reformed along the lines of the laboratory. While the new laboratory sciences had a home in allopathy within a particular network, they were by no means considered the definitive source of medical knowledge by most allopaths, many of whom remained committed to bedside empiricism. Bacteriological reformers still had to persuade reluctant regulars to elevate the status of laboratory knowledge.

Fortunately for bacteriological reformers, the intellectual environment in the United States had become more amenable to assertions of expert knowledge. The Civil War fundamentally upended the intellectual current of the country, especially in the North (Menand 2001). Post–Civil War intellectuals attributed the violent sundering of the Union to the overwrought moralism, inflexible ideologies, and political excesses of the antebellum period. According to thinkers who subscribed to the loose school of pragmatism, ideas had been taken too seriously and proved too explosive. Pragmatists, wanting to avoid rigid ideology, focused their attention on procedures, stressing the importance of expertise, disinterested inquiry, and detached professionalism as the means for a functioning democracy (Menand 2001). Of course, these ideas did not emerge from scratch. Indeed, the early sanitary reformers rallied for disinterested inquiry as superior to urban politics. What pragmatists supplied was a philosophical justification for the role of expertise in democracy, thus tempering the democratic condemnations of hierarchies in knowing, prevalent during the Jacksonian period, and paving the way for the Progressive Era. Emotionally wrought appeals to mass constituencies were being replaced by a buttoned-up model of the scientific expert, and the context of the epistemic contest was altered in the process.

This chapter describes how bacteriological reformers, embedded in this context of larger intellectual change, embraced a new strategy that effectively circumvented the state to achieve professional authority under the epistemology of the laboratory. By enlisting the support of private philanthropies, especially the Rockefeller Foundation, they repositioned the laboratory, moving it to the center of medical knowledge, to promote a new vision of medical science outside the auspices of public institutions. The significant economic resources supplied by Rockefeller allowed a handful of bacteriologists to carry out a multipronged, coordinated reform of America medicine, creating an entirely new medical system that orbited around the laboratory. In other words, with Rockefeller money, reformers transformed the laboratory into an "obligatory passage point" (Latour 1987, 132) through which all medical knowledge on cholera (or any other disease) would have to pass, much as Shakespeare envisioned.

Winning important philanthropic allies, bacteriological reformers remade the epistemological foundation for medicine, created a new identity as scientists for physicians, and dramatically altered the organizational infrastructure of medicine. As gatekeepers of the lab, bacteriologists controlled medical practice. The laboratory became so pivotal that contesting it meant challenging an entire institutional infrastructure of medicine—a task that proved too daunting for any challengers to accomplish. This feat of reorganization was even more impressive given that it was accomplished outside the bounds of the state, as none of the reforms were legally mandated until well after they were in place. By the time the dust of the epistemic contest had settled, allopaths had achieved "one of the more striking instances of collective mobility in recent history" (Starr 1982, 79). Successfully circumventing the type of democratic debate that had long stymied their efforts to professionalize, they overcame the state's traditional commitment to democratized knowing, replacing it with an insulated system of medical expertise in which their epistemic authority went unchallenged. In turn, they created a medical profession unique to the developed world.

THE EPISTEMOLOGY OF THE LAB

Reorganizing medicine around the laboratory required a radical reformulation of the epistemological foundation of allopathic medicine. Bacteriological reformers, recognizing that this would be a difficult sell to many of their peers, attempted to downplay the radicalism of their vision. As discussed in chapter 4, they initially framed the new epistemology of the lab as a continuation of, and improvement upon, radical empiricism, as a *rational* approach to empiricism. Thus, while the laboratory may have been a subversion of radical empiricism (Warner 1991), it was not readily identifiable as such. According to its advocates, bacteriology was simply taking up the radical empiricist mantle of eliminating speculation; advances in bacteriology promised "an epoch in the attitude of medical thought, as well as in medical science, by tending to do away with isms, schools, and theories, and all medical philosophy not founded on demonstrable facts" (Roe 1899, 57). As did radical empiricism, the laboratory rejected speculative theory, championing the role of facts in the production of medical knowledge. But it would bring order to radical empiricism. In a somewhat dialectical fashion, it sought to mold empiricism with a rational underpinning in order to achieve a cohesive program of medical research and avoid the unwieldy, proliferating observations that plagued radical empiricism. What the new epistemology promised was organized and disciplined observation. Framed merely as a technical reform for observation, the laboratory's revolutionary epistemic implications were downplayed. But they remained nonetheless.

Despite the best attempts to de-emphasize the epistemological shift the laboratory demanded, bacteriologists' understanding of empirical observation diverged widely from radical empiricism. Whereas bedside empiricism, as practiced in the United States, was a passive epistemology based on the observations of sensory experience, the laboratory subscribed to an interventionist epistemology. This represented a dramatic departure in the understanding of how medical knowledge could be achieved. Previously, medical facts could be "read" from nature. They were visible and apparent to the careful observer. Now, they needed to be wrested from nature, as they were hidden and obscured. Diseases could not just be perceived; they had to be acted upon to be made visible. In other words, laboratory science stresses *intervention* in natural processes (Hacking 1983). Its "epistemically advantageous" character derives from its ability to enculturate an object, to extract it from nature and then alter it in such a way as to wring scientific knowledge from it (Knorr-Cetina 1999, 27). "Objects are not fixed entities that have to be taken 'as they are' or left by themselves," Knorr-Cetina (1999, 26–27) elaborates. "In fact, one rarely works in laboratories with objects as they occur in nature." Laboratories allow for the control of the complexity of nature by creating an enhanced environment that improves upon nature (Knorr-Cetina 1999) and allows for the emergence of "pure, isolated, phenomena" (Hacking 1983, 226). This is precisely what Koch's culture method for growing bacteria offered. By removing cholera from its natural habitat and placing it in the laboratory, Koch extracted the comma bacillus from its complex natural habitat, where it was essentially invisible and brought it into the lab, where it became visible in its pure form through the microscope, reproducible on cultures, and generally manipulable. By enhancing perception and, in turn, constituting new perceptual objects (Shapin and Schaffer 1985, 36), the lab subjected cholera to spatial and temporal discipline, reducing the "noise" of the disease and rearticulating it as an isolated bacillus. The whole notion of cholera as a germ was dependent upon this special space in which it could be seen as such. Under the epistemology of the laboratory, the researcher did not have to bow to nature's whims to find knowledge; nature could be bent to the will of the researcher. It was this new ideology of scientific expertise based on intervention that was the truly revolutionary aspect of the lab (Maulitz 1979, 92).

In part, bacteriologists' attempt to obscure the epistemological implications of the laboratory was facilitated by the laboratory itself. Sociologists of science have carried out numerous studies of the scientific laboratory for over two decades (e.g., Knorr-Cetina 1999; Latour and Woolgar 1986; Pickering 1984). And while the modern laboratory, operating under "Big Science," with its extensive scope, collective team-based organizational structure, and complex technological apparatus, is much different from the modest bacteriological laboratories of the nineteenth century, the underlying epistemology of the lab remains consistent, if dramatically extended. We can draw on this contemporary research to understand how the laboratory was able to mask the radical demands that its epistemology was placing on rank-and-file allopaths. Sociologists of science have noted the penchant of the laboratory for inducing a particular form of forgetting. Once the lab produces a fact, the intermediary steps are forgotten, purged from the final inscription (Latour and Woolgar 1986). In this process of forgetting, the theoretical rationale (e.g., the bacteriological theory of disease) for performing the experiment in the first place is consumed by the fact (e.g., cholera as a microbe) as are the complex contingencies involved in scientific research. Facts therefore seem to emerge from the lab, rather than being produced by them. Washing away their history, laboratories give facts an empirical veneer that naturalizes them. The laboratory's forgetfulness helped bacteriologists reconcile their new epistemological system with radical empiricism, allowing them to downplay the theoretical commitment to germ theory. Stressing the facts the lab had discovered, they framed laboratory science as *atheoretical*: "Almost from the beginning the student of today is taught methods, where a hundred years ago he was taught theories" (Osler in Thayer 1969, 131). This obscured the revolutionary aspect of the laboratory and allowed for a certain resonance with radical empiricism. The lab was framed not as the embrace of a new epistemology, with its concomitant celebration of certain facts over others (laboratory data over bedside observation), but as a new powerful form of empiricism. Facts were remembered and celebrated; the theoretical and epistemological commitments underwriting these facts were forgotten.

In promoting the epistemology of the lab as a "new era in the history of medicine" (Welch 1920a, 27), bacteriologists sought to elevate technical expertise over clinical experience. Whereas previously a physician's epistemic authority was rooted in his ability to *observe* well, it would now be rooted in his ability to *intervene* well by applying certain technical methods (Warner 1991). Doctors should be granted epistemic authority not because they possessed some art or intuitive skill of observation, but because they possessed the superior technical, scientific knowledge that allowed them to intervene via laboratory analyses. The revolutionary aspect of this shift cannot be understated. Bacteriological reformers were asking their peers to discard decades' worth of medical common sense and to change their thinking about many basic facets of medical practice:

- *Physician Identity*. The epistemology of the laboratory required a shift in the identity of physicians from practitioners of the medical arts to scientists involved with, or at least familiar with, laboratory sciences. As medical knowledge shifted from sorting empirical observations to engineering problems that stressed exactness and precision (Warner 1991), doctors had to be reconstituted as new "epistemic subjects" (Knorr-Cetina 1999). Diagnosis, once deemed a nuanced art of the skilled physician, would now be taken care of by the laboratory, for "no physician who has not had a good experience in the pathological laboratory can be a good diagnostician" (Osler in Thayer 1969, 231). The shift toward seeing themselves as scientists did not go uncontested within allopathy, as evident in the full-time controversy between clinicians and bacteriologists discussed below.
- *Medical Education*. The reforms also suggested an entire new medical pedagogy. Medical education would no longer be transmitted through didactic lectures. All facets of medical education would now emanate outward from the laboratory and promote learning through intervening. Students had to be trained as scientists who could ask questions of nature through laboratory analysis (Borrell 1987). Indeed, education reforms would be the primary means by which bacteriologists carried out their revolution, as they replaced proprietary medical schools with scientific medical schools attached to major universities.
- *The Role of the Hospital*. With the laboratory established as paramount, physicians required access to labs in order to practice medicine. Because labs were expensive, it made economic sense to pool resources and establish them in some centralized place. Hospitals offered such a place. For the laboratory revolution to happen, the locus of medical practice would have to shift from the home to the hospital, which had access to labora-

tory facilities and technical laboratory workers. Traditionally governed by lay trustees and understood as charitable enterprises (Rosenberg 1987a), hospitals needed to be reorganized around the lab, and in the process, elevated from their marginal role of treating the poor to *the* center of medical practice.

- *The Identity of Disease.* As we have seen in its dealings with cholera, the laboratory promised to radically transform the identity of infectious diseases (Cunningham 2002). In the aftermath of Koch's announcement, bacteriologists began a program to have infectious diseases exclusively identified, and defined, by the laboratory. All organizations dealing with disease would be given a laboratory. Furthermore, they would have to become appendages to the lab by engaging in germ hunting. This transformation also would not go unchallenged, no more so than by public health officials, for whom the new identity of cholera clashed with the traditional (and successful) view of cholera as filth.
- *The Patient/Doctor Relationship.* In conjunction with the changing identity of disease, the clinical interaction between patients and doctors was reimagined. This was a subtle, but significant change. Patients' experiential knowledge would no longer have the same weight shaping understandings of disease. With new technologies of observation, physicians would become less dependent on patient testimony in making diagnoses (Rosenberg 2002). This resulted in the "disappearance of the sick man from medicine" (Jewson 1976), as the social distance between the doctor and the patient increased to such an extent that the patient forfeited nearly all control over defining the disease to expert opinion (Katz 2002). The public lacked the technical skill to participate in the conversations over disease. Not only did this shift go against the traditional doctor/patient interaction, it also stood in stark contrast to the democratized epistemologies of other medical sects that recognized patients as active participants in producing knowledge.

The extent of the changes demanded by the laboratory appealed to the professional aspirations of bacteriologists, as they allowed allopaths to reclaim some of their professional mystery (Warner 2002). The laboratory was a restricted space, open and legible only to those who possessed the requisite expertise. Giving medicine back some sense of its "legitimate complexity" (Starr 1982, 59), the laboratory enabled bacteriologists to claim scientific superiority, professional authority, and epistemic privilege. Medical knowledge, once relatively accessible to outsiders, would now be confined to restricted laboratories controlled by allopathic physicians.

ROCKEFELLER MEDICINE MEN

There were two obstacles facing bacteriological reformers in carrying out their ambitious, multidimensional reform. The first was economic: laboratories were expensive, as were the broad, diverse reforms needed to position the laboratory at the center of medicine. Reformers needed access to tremendous resources. This economic problem was compounded by a political one. The natural place to look for such resources was the government. However, the bacteriologists' reform program stood in opposition to the ideals of democracy endorsed by government institutions. The epistemology of the laboratory offered an inherently elitist epistemological vision for medicine. The laboratory sought to restrict, rather than promote, transparency and debate, making medical knowledge the sole province of the allopathic profession. Legislatures had repeatedly refused to regulate the medical market, valuing openness and debate over medical issues as the way to ensure that the best knowledge was achieved. Bacteriological reformers knew this and were skeptical that the government offered a viable partner in remaking American medicine. Instead, bacteriologists sought allies outside of the government who could provide the requisite resources to carry out such an ambitious program. They found an unlikely ally in John D. Rockefeller Sr., American industrialist and founder and chairman of Standard Oil. Rather than convince a wide electorate (or their representatives) of the legitimacy of epistemic authority, bacteriologists were able to persuade a few well-placed, well-heeled elites, like Rockefeller, to support a program to radically remake American medicine.

The last decades of the nineteenth century witnessed a profound transformation in the material and social conditions of the country. Industrial growth after the Civil War resulted in the rise of industrial giants, huge corporations, and large-scale bureaucratic operations (Tratchtenberg 2007). Giant corporations transformed the economic landscape, yielding a stunning array of consumer goods, millions of additional jobs, and ever more wealth concentrated into fewer hands (Diner 1998). And almost no one was as adept as Rockefeller in turning this new economic environment to advantage, accumulating tremendous wealth and economic clout in the process.

It is a great irony of history that Rockefeller would become the catalytic force behind the emergence of modern scientific medicine, given his own "quackish" commitments. Rockefeller's grandmother, Lucy, known locally as a healer of sorts, dispensed herbal remedies and home-brewed concoctions;[3] his father was a self-styled, itinerant "botanic doctor," a throwback in the rural Thomsonian mold. As for Rockefeller himself, he remained a devoted patient to homeopathy his entire life. Dr. Henry Biggar, a close confidant and family homeopathic doctor, treated Rockefeller throughout his life and repeatedly petitioned the wealthy industrialist to support homeopathic causes, which Rockefeller often did. In 1875, Rockefeller invested in a short-lived sanatorium specializing in homeopathic medicine and water cures (Chernow 1998, 183). Even after the establishment of the Rockefeller Institute for Medical Research (RIMR), Rockefeller sent strongly worded letters to its directors demanding that homeopathy be given equal support. Despite backing scientific medicine, Rockefeller remained "notably suspicious when it came to the medical profession" (Chernow 1998, 506); not only did he eschew allopathic treatments, he only stepped foot into the state-of-the-art facilities of the RIMR once in his entire life (Chernow 1998, 475).

Given Rockefeller's unorthodox medical views, the fact that he bankrolled the allopathic medical reforms—reforms that did more than anything to kill off homeopathy—has puzzled historians. Some have dismissed Rockefeller's medical philanthropy as a cynical public relations stunt meant to distract from his controversial labor practices. In a proto-Marxist argument, E. Richard Brown (1979) argues that Rockefeller's capitalist ideology squared nicely with the new medical science, for both ignored the social causes of disease, choosing instead to see social and economic problems as technical problems of engineering. According to Brown, the philanthropic work allowed capitalists to portray disease, not poverty, as the root of misery. While it is certainly true that Rockefeller was not acting *against* his capitalist interests in promoting medical science, historians tend to overstate the degree to which the medical reforms were driven by base economic calculations. For one, Rockefeller was loath to take public credit for or even announce his philanthropic work, failing to take full advantage of the potential public relations coup at his disposal (Chernow 1998). Also, Rockefeller erected a rigid boundary between his philanthropies and Standard Oil. The two enterprises were insulated from one another organizationally, and Rockefeller assumed a generous hands-off stance toward his philanthropies. He left their management to others. Thus, popular depictions in the press at the time notwithstanding, Rockefeller was never a puppet master, manipulating his philanthropic giving with one hand to promote his capitalist empire with the other.

Instead, Rockefeller's philanthropic interest in medicine, and specifically bacteriology, is better understood as emanating from his own use of the laboratory in gaining competitive edge. Often, research in science studies neglects corporate science and, by focusing solely on public science, obscures the long-standing mutual influence between the two (Penders et al. 2009; Shapin 2008). Private sponsorship of public science (like that of the Rockefeller Foundation) is dismissed as driven by base economic motives, rather than reflecting a shared epistemic orientation toward knowledge production. The false dichotomy between *pure* science and more applied research obscures the important role of private research in the development and promotion of science generally. Rather than denigrate the research of Standard Oil as somehow corrupted, it is important to recognize the shared affinity toward the laboratory between bacteriological reformers and Rockefeller's Standard Oil. It is this epistemological affinity that explains Rockefeller's involvement with bacteriology. Just as the democratized epistemologies of homeopathy and Thomsonism resonated with the epistemic cultures of the state legislatures, so too did bacteriologists' epistemology of the laboratory resonate with the epistemic culture of Standard Oil. Once we acknowledge the integral role of the laboratory in industry during this period, industrialists' interest in promoting laboratories in other fields seems less puzzling and, indeed, less cynical. Both Rockefeller and bacteriologists shared a vision of the rationalized production of knowledge by elite experts in laboratories.

In building Standard Oil—"the very modern symbiosis of business acumen and scientific ingenuity" (Chernow 1998, 74)—Rockefeller created an organization that promoted technical innovation as good business strategy. He subscribed to the views of Arthur D. Little, a chemist and early promoter of industrial research, who claimed that "the laboratory has become a prime mover in the machinery of civilization . . . for research is the mother of industry" (quoted in Shapin 2008, 96). Standard Oil cornered the oil industry, not by focusing on the production of crude oil, but through refining. Its profitability depended on scientific, technical acumen to transform crude oil into end products like gasoline, kerosene, and petroleum. Presaging modern research and design departments, Rockefeller created a committee structure of experts within Standard Oil to study the technical problems of oil refining. To increase efficiency and ultimately profit margins, the findings of this cadre of internal scientists were made widely available to the entire company. Rockefeller explained his institutional organization of inquiry: "A company of men, for example, were specialists in manufacture. There were chosen experts, who had daily sessions and study of the problems, new as well as old, constantly arising. The benefit of their research, their study, was available for each of the different concerns whose shares were held by these trustees" (quoted in Chernow 1998, 229). Within Standard Oil, technical experts were given space to innovate, to make improvements either in the refining or manufacturing processes, which would then be disseminated throughout the company. Rockefeller created insulated research spaces for scientific tinkering.

Standard Oil's epistemic culture of innovation was evident in one particular episode in the mid-1880s. From its inception, Standard Oil faced the chronic threat of diminishing oil reserves. With his company dependent solely on Pennsylvania oil, Rockefeller perpetually worried about the reserves' depletion. Some of these fears were assuaged by the 1886 discovery of oil in Ohio and Indiana, until it was realized that this oil burned dirtier than Pennsylvania crude. Undaunted, Rockefeller hired a German-born chemist, Herman Frasch, to turn the "Lima crude" into a marketable commodity. Frasch solved the problem, and for fifteen years, his patents "furnished dazzling profits for Rockefeller and Standard Oil and boosted the status of research scientists throughout the industry" (Chernow 1998, 286–287). When Rockefeller first hired him, Frasch was probably the only trained petroleum chemist in the United States, but his success solidified Rockefeller's commitment to the laboratory. By the time Rockefeller retired, there was a lab in every refinery and even one in Standard Oil's headquarters in downtown Manhattan.

Whether refining oil or researching disease, Rockefeller displayed a commitment to the epistemology of the laboratory, creating organizations that reflected this commitment. Like bacteriologists studying disease, Rockefeller was trying to tame nature by intervening in natural processes via the laboratory. It was this epistemic resonance, more than class interest, that explains how Rockefeller, a man committed to homeopathy, became "the financial father" (Gates 1911a, 2) of scientific medicine.

Creating a Pure Laboratory

Before the federal government, under President Theodore Roosevelt, busted industrial trusts, corporate behemoths like Standard Oil had accumulated wealth to previously unfathomable levels. Unable to recirculate their profits, the new wealthy elites like Rockefeller, Andrew Carnegie, and Andrew Mellon turned to philanthropy. Part of the impetus for philanthropic giving came from an elitist sense of responsibility, best expressed in Carnegie's *Gospel of Wealth* (1889), which sought to manage wealth and charity in an efficient way so as to benefit society on the whole. But part of it came from a particular ideological hostility toward the rough-and-tumble democratic politics of the Jacksonian period that arose in the Progressive Era. Decrying the poor policy outcomes of egalitarian democracy, the Progressive movement enlisted experts to rationalize the emergent social order (Wheatley 1988). Philanthropy shared many of the same aims as Progressivism, and by 1890, philanthropy was beginning to be seen as a solution to the institutional underdevelopment resulting from "a nation born in a day" (Wheatley 1988, 16). Weary of mass democratic politics, philanthropists sought to circumvent the vagaries of party politics by promoting sober professional expertise as the means to better policy (Wheatley 1988).

The Rockefeller Foundation became the very model of modern philanthropy, deploying its resources to support the rational application of expert knowledge. Demonstrating his hands-off approach to philanthropy, Rockefeller deferred much of his philanthropic decisions first to Frederick T. Gates, Rockefeller's appointed philanthropic adviser—the "tutelary spirit of the Rockefeller philanthropies" (Chernow 1998, 470)—and later to his son, John D. Rockefeller Jr. Besieged by requests for gifts, Rockefeller turned to Gates to manage his philanthropy in 1892, and for the next twenty years, Gates served as the broker between Rockefeller and Welch's network of bacteriologists. More so than Rockefeller, it was Gates, Rockefeller Jr., and later Abraham Flexner who translated a general epistemological resonance into a radical program of medical reform.

Whereas Rockefeller shared only a vague commitment to the laboratory with bacteriologists, Gates developed a specific commitment to bacteriology as the future of medical science. After an illness in the spring of 1897, Gates read William Osler's *The Principles and Practice of Medicine* (1895) while on vacation in the Catskill Mountains (Markel 2008). Although impressed with the detailed nature of the book that described the new science of bacteriology, Gates was shocked by the backwardness of American medicine, which boasted cures for only a handful of ailments. As he later recalled:

> I found further that a large number of the most common diseases, especially of the young and middle aged, were simply infectious or contagious,

> were caused by infinitesimal germs. . . . I learned that of these germs, only a very few had been identified and isolated. I made a list, and it was a very long one at that time, much longer than it is now, of the germs which we might reasonably hope to discover but which as yet had never been, with certainty, identified, and I made a very much longer list of the infectious or contagious diseases for which there had been as of yet no specific found. (quoted in Corner 1964, 579)

List in hand, Gates cut his vacation short, returning immediately to his Manhattan office with a vision of Rockefeller-funded medical science germinating in his mind (Markel 2008). He was convinced that the scientific study of medicine, "woefully neglected in all civilized countries and perhaps most of all in this country" (quoted in Corner 1964, 579) was on the precipice of dramatic breakthroughs. It just lacked the proper support. In Gates's budding vision, Rockefeller could provide the support to transform medicine, just as he had the oil business, as "the precise analysis of the human body into its component parts is analogous to the industrial organization of production" (Brown 1979, 119). Just as laboratory analysis solved technical problems of industrial production, so too could it heal the technical problems of the body.

The first step was to create an independent laboratory dedicated solely to primary medical research and shielded from all other concerns. The plan was innovative and daring. And to most within the profession it was foolhardy, for it "seemed quite rash, even quixotic to pay grown men to daydream and come up with useful discoveries" (Chernow 1998, 471). Few beyond those involved recognized the scientific, professional, and epistemic significance that this new laboratory would have on medicine. After gaining support from John D. Rockefeller Jr., Gates sent lawyer and Rockefeller advisor, Starr Murphy to Europe to study the Pasteur Institute and Koch Imperial Health Office, the world's foremost bacteriological laboratories. He also met with two reform-minded physicians, Emmett Holt and Christian Herter, to solicit names of those who could help shape the institute. The doctors directed him to none other than their former teacher and dean of the Johns Hopkins Medical School, William Welch. Thus began a relationship between Welch and the Rockefeller philanthropies that would last three decades.

The inclusion of Welch linked Rockefeller money to Welch's network of bacteriologists. As a chief adviser in the search to staff the new institute,

Welch sought to extend his program of bacteriological reforms that he had already begun at Johns Hopkins. Welch, who had long recognized that "large endowments are necessary for laboratories especially, and here in the Eastern States at least we must look to private philanthropy for this purpose" (Welch 1920b, 45), now had access to the resources necessary for his program. Assembling a team of a veritable who's who of American bacteriology that included Hermann Biggs, Simon Flexner, Christian Herter, T. Mitchell Prudden, and Theobald Smith, Welch played an instrumental role in the founding the Rockefeller Institute for Medical Research (RIMR), the first independent medical laboratory in the country, in 1901. Unattached to any municipal organization or educational institution, the RIMR was intended as a place where researchers could pursue basic medical research without any competing commitments. This independence was jealously guarded, as Welch believed that the RIMR would only be successful if researchers were free of distractions to explore what they wished. Agreeing, the editors of the *Medical Record* ("Rockefeller's Institution for Medical Research" 1901, 907) predicted that the RIMR "will set free the men of the American medical profession educated in scientific lore, and will permit them to follow the bent of their minds, to the honor of their country and to the good of mankind, untrammeled by sordid considerations." This independence was codified in the laboratory's bylaws, which gave the scientists unlimited control over the research agenda. Welch had secured an institutional stronghold that would serve only the dictates of science.

While there was consensus among those involved as to the mission of the RIMR, Rockefeller himself did raise concerns periodically. Prodded by his friend Dr. Biggar, he was especially concerned about the RIMR's exclusion of homeopathic research. To mollify Rockefeller, Gates drew on a common trope of the medical reformers, appealing to the theoretical neutrality of science. The new science did not seek to replace homeopathy with allopathy; it sought to transcend sectarian medicine altogether. The RIMR was an institution that was "neither allopath nor homeopath, but simply scientific in its investigations into medical science" (Gates in Corner 1964, 582). Medical science was the future, sectarianism the past. Gates stressed the empiricism of discoveries—and drew on the amnesia induced by the lab—to convince Rockefeller that his money was being spent in an agnostic fashion. This was a bit disingenuous, as Gates was a firm critic of homeopathy, deriding Hahnemann as "little less than a lunatic" whose system's popularity was based on the "ignorance and credulity of . . . patients" (Gates in Corner 1964, 577).

He even went so far as to compose a series of detailed memos to Rockefeller, in response to Biggar, critiquing homeopathy. Still, while Rockefeller continued to voice concerns, his age,[4] his faith in Gates, and his respect for the autonomy of experts prevented him from intervening any further on behalf of homeopathy.

With the establishment of the RIMR, bacteriologists found themselves in a position of authority unthinkable just a decade earlier when they were exiled to a handful of woefully underfunded laboratories. Not only did the RIMR offer a purified epistemic space under control of bacteriologists; with Rockefeller's stamp of approval, bacteriologists had access to nearly unlimited resources. And aspiring medical scientists now had career prospects. As medical research became more lucrative and prestigious, they began to seek careers in the laboratory. These ambitious young medical scientists were not content to stay in the RIMR, and the institute became "an incubator for a group which aspired to lead in reforming medicine and medical education" (Wheatley 1988, 39). As the network of bacteriologists grew and dispersed, they brought their influence, and their program to remake medicine around the laboratory, elsewhere.

REMODELING AMERICAN MEDICAL EDUCATION

One lab, however well endowed, does not an epistemological revolution make. To take hold, a new generation would have to be socialized into laboratory science. The epistemological change bacteriologists sought needed pedagogical reforms. Throughout the nineteenth century, U.S. medical education experienced a race to the bottom as proprietary schools lowered standards in an attempt to attract more students and increase profits. Students would graduate without having attended a birth, witnessing an operation, and often without even examining a patient (Ludmerer 1985, 12). When Charles Eliot became president of Harvard in 1869, his proposal to require written examinations for graduation was met with resistance from the director of the medical school who asserted, with little exaggeration, that a majority of his students could hardly write (Burrow 1963, 9). The situation got so bad that, as late as 1887, the Maine State Board of Health had an eight-year-old boy apply to a number of medical schools. More than half accepted him (Duffy 1993, 203).

Despite this race to the bottom, things were not hopeless. In the 1880s, a handful of medical schools began to elevate their standards. These reforms took place in the context of the coming-of-age of American universities (Starr 1982, 112). In the late 1800s, a handful of elite American colleges sought to remodel themselves along the lines of the German university, with its focus on research, graduate education, and the sciences (Banta 1971; Bonner 1963; Starr 1982; Veysey 1965). Under the German model, higher education was centered on *producing* knowledge, rather than merely *conveying* it—the dominant approach of the English universities that had long served as the model for American medical education (Ludmerer 1985). This represented a shift from an education that took theology as its model discipline to one organized around the sciences, reflecting an awareness that the increasing complexity of knowledge could only be addressed by pedagogy that focused on critical thinking rather than rote memorization. Given their strong ties to German institutions and their interventionist epistemology, it is not surprising that bacteriologists embraced the German model as a solution to American medical education (Bonner 1963).

Bacteriologists began to make some real gains along these lines. The most successful of these early efforts were Welch's reforms at Johns Hopkins University—reforms that would serve as the model for the future of medical education in the United States. As with the establishment of the RIMR, the resources for the educational experiment at Johns Hopkins came from private philanthropic giving. In 1873, Johns Hopkins, a merchant and banker, left $7 million upon his death for the establishment of a modern university in Baltimore, the largest philanthropic gift ever bestowed in the United States at the time. In many ways, Hopkins was an ideal place to experiment with medical education. While "the expense of laboratory teaching [had] been urged to bar it" from most medical schools (AMA 1892b, 111), Hopkins's generous endowment provided for the modern trappings of a university, including well-equipped laboratories and elite teacher-investigators, poached from other medical departments, to teach in these labs (Banta 1971). And as a new university, it was able to undertake such reforms with a blank slate (Veysey 1965, 129), rather than fight the endless internecine battles with traditional faculty that delayed similar reforms in universities like Harvard.

In 1894, President Daniel Coit Gilman hired Welch to be the dean of a medical school that did not yet exist, giving him the freedom to create the school as he saw fit. Arguing that "the proper teaching of medicine now requires hospitals, many laboratories with expensive equipment and a large force of teachers, some of whom must be paid enough to enable them to devote their whole time to teaching and investigating" (Welch 1920b, 46),

Welch envisioned an institution centered on the laboratory sciences. His philosophy for the school—"We hold that the medical arts should rest upon a thorough training in the medical sciences, and that, other things being equal, he is the best practitioner who has that thorough training" (quoted in Flexner and Flexner 1941, 223)—represented a significant pedagogical innovation for the period. The laboratory's "great service is in developing the scientific sprit and in imparting a living, abiding knowledge, which cannot be gained merely by reading or being told about things," argued Welch (1920a, 71). "So important are these ends, that it seems difficult to overestimate the value of the laboratory in scientific teaching." As such, the school's pedagogy embodied the new interventionist epistemology of the laboratory and reproduced it via the socialization of the next generation of elite physicians by encouraging them to participate in research. The student "no longer merely watches, listens, memories; he *does*" (Flexner 1910, 53). Learning medicine by *doing* research, students became better equipped to adapt the new sciences to the practice of medicine: "The knowledge derived from actually seeing, touching, experimenting, is of course more real and impressive than that which comes simply from reading and from listening to lectures" (Welch 1920a, 57). Manipulating disease in the laboratories, students learned to deal with problems scientifically, whether in conducting research or treating patients. And like the RIMR, Hopkins created a pure research environment, where a student or faculty member experienced "freedom from the cares of the world, liberty to pursue the search for truth in his own way, liberty of thought, liberty of utterance" (Osler in Thayer 1969, 305).[5]

Still, Hopkins was a newcomer in the world of American higher education. Its transformation from an experiment to *the* model of medical education involved a concerted campaign, carried out through the AMA by bacteriological reformers with ties to Hopkins and access to philanthropic funding. Of key cultural import was the Flexner Report published in 1910.[6] In 1904, as part of a general internal reorganization, the AMA established the Council on Medical Education (CME) to reinvigorate its program of educational reforms. Dominated by bacteriologists, the CME decided to carry out an investigation of medical schools, assessing all medical schools in comparison to the laboratory education of Hopkins. Because of the internal resistance to reforms among proprietary school faculty, the CME sought assistance from the Carnegie Foundation to produce an unbiased "outsider" report. The independence of the Carnegie Report, however, was little more than formal window dressing; from the beginning, it was acknowledged that study would be done in conjunction with the CME, as the CME would furnish much of the research and data collection (Berliner 1985; Burrow 1963). To conduct the survey, the Carnegie Foundation chose Abraham Flexner, an educational reformer whose brother was Simon Flexner, a Welch student and future director of the RIMR. As a graduate of Johns Hopkins himself and "a great admirer of William Welch" (Duffy 1993, 208), Flexner, like his brother, was firmly entrenched in Welch's network of bacteriologists. He saw Hopkins as the standard by which all other medical schools should be judged, as it "was the first medical school in America of genuine university type, with something approaching adequate endowment, well equipped laboratories conducted by modern teachers, devoting themselves unreservedly to medical investigation and instruction, and with its own hospital, in which the training of physicians and the healing of the sick harmoniously combine" (Flexner 1910, 12). So he went about judging schools according to these criteria.

The report itself was the prototype of the agenda-setting surveys that would become a common tool for reformers throughout the Progressive Era (Wheatley 1988). Flexner assessed medical schools along a number of dimensions: (1) entrance requirements, (2) size and training of faculty, (3) nature and extent of endowment, (4) adequacy of labs and lab teaching personal, and (5) availability of clinical resources and nature of clinical appointments. While each dimension was relevant, Flexner made it explicit that adequate laboratories would trump all other factors. If a school lacked adequate facilities for laboratory science, it was declared inferior. Legitimate medical knowledge was equated with laboratory science, and in turn, medical education should revolve around the lab: "For purposes of convenience, the medical curriculum may be divided into two parts, according as the work is carried on mainly in laboratories or mainly in the hospital; but the distinction is only superficial, for the hospital is itself in the fullest sense a laboratory" (Flexner 1910, 57).

In all Flexner visited 155 medical schools, 32 of which were affiliated with alternative sects (Bordley and Harvey 1976). His conclusions were damning: "It is a singular fact that the organization of medical education in this country has hitherto been such as not only to commercialize the process of education itself, but also to obscure in the minds of the public any discrimination between the well trained physician and the physician who has had no adequate training whatsoever" (Flexner 1910, x). Flexner recommended increased entrance standards, a four-year curriculum, and a drastic reduction in the number of medical schools to 31. Overall, the report proposed a wholesale transformation of medical education by which physicians would be trained in the epistemology of the laboratory to become "scientists in terms of treating each new clinical encounter as an exercise in scientific inquiry" (Flexner 1910, 9).

The report garnered much media attention, as it brought to light some egregious inadequacies of many medical schools. And though there was resistance to the report by many schools—F. W. Hamilton, president of Tufts University, argued that Flexner "had adopted certain arbitrary standards as to methods . . . which may be interesting to him but is worthless for anybody else" (quoted in Wheatley 1988, 51)—the AMA made effective use of this public outcry. The CME adopted a recurring system of ranking schools based on Flexner's criteria, ensuring the continued saliency of the report. The CME exerted continuous pressure on medical schools by only recognizing graduates from schools that met these criteria, and later using their influence on licensing boards to essentially legalize these standards. Through this continuous pressure and its later cozy relationship with licensing boards, the CME became, de facto, "a national accrediting agency for medical schools, as an increasing number of states adopted its judgments of unacceptable institutions" (Starr 1982, 121).

While the report focused public attention on medical education and elevated the cultural cachet of Johns Hopkins, its most lasting significance was in capturing philanthropic attention (Duffy 1993). Much of the reforms were less the result of the Flexner Report and more of the carrot offered by foundations (Stevens 1971). To carry out the proposed reforms—"an enormously expensive affair" (Welch 1920b, 59)—reformers needed copious institutional support. Once again, Rockefeller money was paramount. Gates devoured the Flexner Report, viewing it as a road map for future Rockefeller giving. When he invited the author to lunch to discuss it, Flexner pointed to two maps in his book—one showing the locations of the medical schools he visited, the other showing what the country needed. " 'How much would it cost to convert the first map into the second?' Gates asked and Flexner replied, 'It might cost a billion dollars.' 'Alright,' Gates announced, 'we've got the money, come down here and we'll give it to you' " (quoted in Chernow 1998, 492). Later, when Gates posed a hypothetical question to Flexner about the best way to invest $1 million in medical science, Flexner replied that he would give the money to Welch to do what he liked with it. Gates agreed, further consecrating Welch's vision at Johns Hopkins "as the prototype to be emulated by recipients of Rockefeller money" (Chernow 1998, 492). With Gates support, the Rockefeller philanthropies used the Flexner Report to guide their giving, a policy formally institutionalized when Rockefeller invited Flexner to serve on the foundation's General Education Board (GEB).

In the end, Flexner provided the blueprint; Gates provided the financial resources and incentives to make Flexner's suggestions a reality. Controlling the allocation of Rockefeller funds, Gates and Flexner, in consultation with Welch, were able to choose the winners and losers in medical education. And education reform was the chief mechanism by which doctors established bacteriology as a key piece in the professionalization of medicine. For Flexner and his colleagues, rationalizing medical education meant standardizing it as an education focused on scientific research in the laboratory. Previously, there were many possible routes to becoming a doctor; now there was only one—through the laboratory. Reformers did not have to choose a one-size-fits-all model of education; indeed, many clinicians clamored for a medical educational system with multiple tiers (Ludmerer 1985). But Flexner was insistent; an acceptable medical education needed to look like that of Hopkins. By effectively forcing medical schools to adopt the scientific model of medical education in order to receive philanthropic funding, educational reformers dashed any dreams of diversity in medical training. Doctors were to become scientists. Medicine was to have a single epistemological foundation.

These reforms were realized not by popular fiat among the rank and file of the profession—indeed many clinicians disparaged the reforms—nor by convincing state legislatures to back them. Instead, elite bacteriological reformers built a new model of medical education by convincing a few philanthropies to bankroll their program.

EPISTEMIC CLOSURE

While medical education was central because it socialized future generations into the epistemology of the laboratory, other institutions like hospitals, licensing laws, and boards of health were consolidated as well under the laboratory. This consolidation, along a number of different fronts, was not achieved without struggle. Nevertheless, by 1920, medical reformers had achieved epistemic closure by radically remaking the organization of American medicine along the lines of the laboratory. As E. O. Shakespeare imagined with cholera, all diseases would be filtered through the laboratory, which would identify, understand, and then finally eliminate them.

Hospitals and the Full-Time Controversy

During the nineteenth century, the hospital was a marginal institution established to serve the poor (Duffy 1993; Rosen and Rosenberg 1983; Rosenberg 1977, 1987; Rosner 1982; Rothman 1991; Starr 1982). Given the profession's commitment to bedside empiricism, the locus of medical practice was the patient's home. Yet as the new epistemology of the laboratory reallocated the valuation of evidence away from idiosyncratic bedside observations to laboratory tests, doctors increasingly needed access to labs to practice medicine. Because the costs of establishing laboratories were prohibitive for all but the most elite practitioners, hospitals became a central location where doctors could access the services of the lab:[7] "There ought to be, there must be laboratory facilities in and directly connected with every modern hospital. It requires no demonstration that rational treatment is not possible without a correct and minute diagnosis" (Jacobi 1897, 114). With the ascendency of the laboratory, hospitals assumed a prominent place in medicine, morphing from locally based charitable institutions, under the control of lay trustees, into large-scale bureaucracies committed to medical science (Duffy 1993: Rosenberg 1987a; Rosner 1982).

This reorganization of the hospital was inextricably tied to educational reforms. One of Welch's innovations at Hopkins was to integrate the hospital into medical education. Welch viewed the hospital, and the clinical work that went on there, as an extension of the lab: "The teaching of the clinical subjects should be carried out along the same general lines as those of the laboratory" (Welch 1920b, 127). In his report, Flexner codified this view in establishing standards for medical education, going so far as to define the hospital as a lab itself (Flexner 1910, 57). When he joined Rockefeller's GEB, Flexner sought to extend this Hopkins model to all medical schools by requiring clinicians who taught in medical schools to become full-time faculty, to renounce their private practices, and to focus on research and teaching. From the start, faculty who taught the basic laboratory sciences were hired on a full-time basis, an employment status that spoke to the primacy of research in their identity as scientists. Clinical faculty, however, had a more ambiguous status with ties both to the school and to their own private practices. Indeed, hospitals had long served an important role in advancing the careers of local elite doctors (Rosenberg 1987a). Flexner viewed this as a conflict of interest that left clinicians susceptible to the corrupting influence of commercialism and careerism. To remedy this situation, he acquired Rockefeller funding for an experimental clinical full-time program at Hopkins. Satisfied with the results there, he planned to extend the program to all medical schools.

Flexner's full-time program unleashed a powerful backlash among clinicians who saw medicine more as an art than a science. They denounced the program as a bald attempt to bring clinicians under the control of laboratory technicians, worrying that it "might remove them from the beside to the bench" (Maulitz 1979, 92). While many clinicians welcomed the new tools offered by the laboratory, they were hesitant to privilege laboratory knowledge over clinical experience. After all, the bulk of regulars still practiced in ways far removed from the new ideals of biomedical research (Katz 2002, 41). It was alright for the lab to dictate the practices of public health officials, but when it came to the practice of medicine, clinicians were to have the final word. Some cautioned the profession against succumbing to "bacteriomania" (Jacobi 1885, 172). Others, like Alfred L. Loomis (1888b, 70), the president of the New York State Medical Society, warned those who were in haste to elevate "the experimental above the practical" of the fetishization of the laboratory. Allopathic physicians,

> cannot safely forsake the rich storehouses of clinical observations that have been gathered by so many master-minds, and withdraw into the recesses of the laboratory; for although in the work of the laboratory we hope to find the solution to many of the problems with which we are now struggling, it must be remembered that the special field of medical investigation is, and ever will be, the study of disease in its activities. (Loomis 1888b, 70)

Noting laboratory science's limited impact on therapeutics, clinicians worried that their pupils would lose perspective in an educational program focused primarily on the laboratory, as they whiled away their precious time "in the labyrinths of Chemistry and Physiology" (Bigelow 1871, 8). William Osler, perhaps the most distinguished clinician in the United States, vehemently opposed the full-time program, lamenting "the evolution throughout the country of a set of clinical prigs, the boundary of whose horizon would be the laboratory, and whose only interest was research" (Wheatley 1988, 69).

Pitted against the well-connected and resource-rich network of bacteriologists, clinicians could not compete with the power of Rockefeller's money. Flexner, refusing to bow to pressure, made the full-time program the centerpiece of the GEB's reforms; medical schools that sought Rockefeller money had to pass a litmus test to receive funds. No full-time program, no money. Medical schools were all but required to adopt it to survive. As more and more university hospitals embraced the full-time program and used the funds to build up their infrastructures, hospitals generally became organized according to the epistemology of the laboratory. In the internal allopathic debate over medicine as an art versus medicine as a science, science won. Once again, Rockefeller finances dictated the winners and losers. Local cliques of clinical physicians were replaced with a national network of clinical scientists, shifting the power in the hospitals from local communities to Welch's network of laboratory scientists (Wheatley 1988). And lay trustees, long the governors of hospitals, were elbowed aside as control went to Flexner's physician/scientists (Rosenberg 1987a).

Hunting Germs

Public health had long frustrated allopathic dreams, offering enticing resources but maintaining an ecumenism that prevented allopathic takeover. After bringing medical education and hospitals under the control of the laboratory, bacteriologists turned their attention to public health. The germ theory sought to reorient public health away from cleaning up filth toward hunting germs. But like clinicians, public health officials were reluctant to give up their traditional practices, especially since the multiple cosmetic sanitary reforms had won them public esteem. Many public health officials balked at embracing bacteriology, so much so that as late as 1915, the uncompromising bacteriologist and health officer of Providence, Rhode Island, Charles V. Chapin (1934b [1915], 37) complained, "There is probably not a single large municipal health department in the country which is operated along strictly logical lines." Reiterating "the timeworn phrase about dirt and disease" (Chapin, 1934c [1902], 23), the boards of health had failed to adopt the targeted "rational" approaches suggested by bacteriology. Chapin complained that "the age of bacteriology has produced an immense amount of scattered information. But no one as yet had attempted to draw it all together into a coherent patterns of logical public health principles and measures" (Chapin quoted in Cassedy 1962, 110), and he vowed to bring public health in line with the laboratory.

While upsetting to bacteriologists, this resistance was not overly disconcerting to most reformers, public health officials like Chapin aside. Having consolidated authority in a number of different organizations, regulars were now less dependent on public health for their prestige. As these other organizations gained strength through the laboratory, bacteriological reformers became less concerned about public health and boards of health. As a result, allopathy's long-felt ambivalence toward public health was allowed full flowering in the twentieth century. Regulars were content to leave sanitation to public officials, as long as these officials were willing to defer to the laboratory in their educational and sanitary programs. The GEB provided funds to establish schools of public health attached to, and controlled by, medical schools, but their graduates were not to be doctors. Organizing public health education in this way, sanitary science became a residual category for allopaths, a mere appendage to the laboratory where the true breakthroughs were happening. Public health was demoted in allopaths' imagination, no longer placed at the forefront of controlling disease, but reduced to the practical application of laboratory findings. This new understanding of public health was reinforced in the Rockefeller Foundation's forays into the field, both domestically and abroad, as it only funded programs in which public health interventions were based on the laboratory sciences.

Rather than fight to make public officials doctors, bacteriologists allowed public health to remain in lay hands, provided that boards of health did not encroach on medical practice. The AMA's strategy toward the boards of health shifted from one aimed at control to one aimed at mitigating their influence and protecting the autonomy of individual physicians. As long as physicians maintained ultimate authority, public health programs would be supported. However, if they started to interfere with the autonomy of the individual practitioner, allopaths ensured their demise. For example, the AMA effectively killed a campaign to develop public health dispensaries, which they viewed as competition (Starr 1982). Thus, while regulars tolerated a degree of ecumenism on the actual boards, they effectively circumscribed their activities.

Homeopathy's Enfeebled Resistance and the Mirage of Cooperation

The consolidation of multiple organizations under bacteriological control did not bode well for homeopathy, which saw its tenuous grasp of medical practice evaporate in the early twentieth century. Having ceded its claims to the germ theory, every gain by the lab was a gain for allopathy at the expense of homeopathy. Some historians have argued that rather than suffer defeat, homeopaths chose to join allopathy and cooperate with the medical reforms (Rothstein 1992; Starr 1982). But this is history from the viewpoint of the winner. While it is true that some homeopaths did convert, most resisted the laboratory and its vision of medicine. This enfeebled resistance, however, could not withstand the multifaceted reform program backed by Rockefeller resources.

Bacteriological reformers offered a more ambiguous target for homeopaths than past allopathic reformers, as they tempered their exclusionary rhetoric. For decades, homeopaths had made great political hay by exposing the exclusionary practices of the AMA, condemning such efforts as elitist and antidemocratic. Whenever regulars sought to gain some special recognition, homeopaths cried foul to state legislatures, arguing that regulars were trying to illegitimately monopolize medical practice and stifle debate. These arguments resonated with the legislatures, resulting in the universal repeal of state licensing laws and homeopathic inclusion in government bodies. Homeopaths were able to maintain this equal legal standing even during the bacteriological revolution. Between 1870 and 1890, regulars sought to reestablish licensing laws at the state level that targeted newer "quacks," like osteopaths and chiropractors (Baker 1984). Resigned to the enduring presence of homeopaths and eclectics, regulars hoped to include them in licensing in a manner that would facilitate ultimate allopathic control, proposing legislation for a single board of medical examiners in which regulars were the majority. Alternative sects countered with legislation proposing either separate boards for each sect or a single board with equal representation. As before, state legislatures sided with alternative medical movements, refusing to grant allopathy control over the boards, adopting one of the two alternative proposals.

These de jure, formal legal protections did not preclude homeopaths from being effectively shut out of medical practice de facto. As allopathic physicians seized control of the organizational infrastructure of medicine through the lab, they used their organizational leverage to ensure that homeopaths remained secondary medical citizens, despite their legal equality. The posture they adopted, however, differed from the vitriolic rhetoric of regulars past. Bacteriological reformers framed their interventions as moving beyond *both* homeopathy and allopathy, transcending the bitter sectarian debates for a medical system committed not to dogma, but to science. Flexner (1910, 156) argued, "Modern medicine has therefore as little sympathy for allopathy as homeopathy. It simply denies outright the relevancy or value of either doctrine." Whereas the "sectarian begins with his mind made up," science "believes slowly; in the absence of crucial demonstration its mien is humble, its hold is light," and in the process, it "brushes aside all historic dogma" (Flexner 1910, 157). Once again, drawing on the forgetfulness of the laboratory, the reformers disavowed theoretical commitments. The future would not be shaped by sectarian theories, but by the "methods of thought and observation and in developing the scientific spirit" (Welch 1920a, 5). Gates justified the actions of the Rockefeller Foundation in similar terms: "The day of dogma and philosophic formulae in science has passed away. . . . Medicine is becoming no longer a creed but a science" (Gates 1911b, 3). Framed as atheoretical and nondogmatic, medical science would lead to the "ebbing vitality of homeopathic schools" in a "striking demonstration of the incompatibility of science and dogma" (Flexner 1910, 161). This more measured tone proved tough for homeopaths to combat, for whoever defended one system over another, whether it was homeopathy, eclecticism, or allopathy, was portrayed by bacteriological reformers as out of touch and retrograde.

This commitment to the atheoretical science of the future was not just a strategic or cynical ploy on the part of the reformers. It resulted in tangible reforms to the way allopaths ran their profession. For decades, the AMA had focused on delineating the boundaries between allopathy and "quackery," by internally policing its members through the Code of Ethics and no consultation clause. In 1903, bacteriological reformers successfully lobbied for a revision of the Code of Ethics. Arguing that the era of sectarianism was over, they stressed the importance of freedom of inquiry and consultation for the development of the new science. The old Code of Ethics was replaced with a new Statement of Principles. Ostensibly, homeopaths were now welcome to join allopathic professional societies, provided they give up their commitment to sectarian dogmas. They would be allowed to practice their therapeutics but not to publicly advocate for, or identify with, the homeopathic system.

Homeopaths struggled to respond to allopathy's new positioning. The AMA's public display of welcome reduced the rhetorical punch of homeopathy's traditional arguments that juxtaposed their openness, transparency, inclusiveness, and commitment to debate with the exclusionary practice of allopathy. Now, at least superficially, regulars were abandoning the very practices that homeopaths had long transformed into legislative achievements and public sympathy. Internally, homeopaths debated what this all meant. How genuine was the AMA's invitation? Could homeopathy have a place in bacteriology? Once again, these debates pitted older homeopaths and veterans of numerous sectarian battles (the "highs") against the younger homeopaths (the "lows"), for whom the battles of the past had less experiential relevance. DeWitt G. Wilcox (1904, 6) voiced the seduction that the new medical science held for the lows: "We are, to my mind, entering an era of new medicine, not the product of any one school, but the logical outgrowth of scientific research along all lines of life and health. . . . Gentlemen, we have got to get into the bandwagon of progress, or walk; and unless we do get in we have no right to complain because the old school holds the reins and pounds the brass drum." Lows believed that homeopaths "need not approach the study of bacteriology with the slightest fear that it will destroy the well-grounded temple into which they have built their hopes and allied their dreams" (Maddux 1892, 1). For their part, highs acknowledged that "the spirit of the times is encouraging amalgamation and co-operation," but they expressed concern that "the dominant school of medicine is not yet ready to accept these conditions" (Wood 1902, 39). Recalling past battles, highs smelled an allopathic trap.

While these internal debates weakened the cohesion of homeopathy, in many ways they were beside the point. By the first decades of the twentieth century, allopathic physicians had positioned themselves organizationally to kill off homeopathy. Having established the laboratory as an obligatory passage point around which all medical organization pivoted, allopathic physicians had gained such a stranglehold on the practice of medicine that they were able to effectively prevent homeopaths from participating in the new organizational infrastructure. The AMA could change its tone because the locus of its power was now centered in the new medical infrastructure it had built using Rockefeller funds.

The marshaling of philanthropic resources to attack homeopaths was most evident in medical education. Flexner "excoriated almost all the existing homeopathic, eclectic, and osteopathic schools for poor standards, and made no allowance for the survival of any of them in his reconstruction plans for medical education" (Gevitz 1992, 84). It followed that his criteria for assessing medical schools was hostile to homeopathic schools. As these assessments determined access to funding, it was no accident that in the twenty-five years following the report, homeopathic schools decreased from fourteen to two, largely due to financial inadequacy. Furthermore, as states adopted and consolidated licensing boards, the boards adopted the AMA's ranking of medical schools, granting licenses only to those educated in institutions that achieved an adequate score. In this way the boards conformed to the ecumenical letter of the law, while institutionalizing standards that penalized homeopathy. The boards standardized basic scientific training for all aspiring doctors, regardless of sect; the only differences in evaluation between sects were restricted to therapeutics. These basic standards promoted laboratory sciences, while demoting the status of pharmacology and symptomology, the central disciplines of homeopathy (Coulter 1973).

Leveraging their new position of organizational authority, the AMA sought legislation in other areas of medicine as well, which mitigated the formal ecumenism of the licensing laws. These included hospital and prescribing/drug regulations and later eligibility for insurance reimbursement (Starr 1982, 333). For example, the Pure Food and Drug Act, endorsed by the AMA, instituted tighter regulations on the production of drugs and thereby prohibited homeopaths from creating their own medications—a central component of their distinct identity (Coulter 1973). Additionally, since allopathic physicians controlled hospitals, homeopaths were forced to conform to their standards in order to practice in the new epicenters of medicine. Unable to lay a claim to the new medical science of bacteriology and compelled to join allopathy lest it be shut out from medical practice, homeopathy was increasingly marginalized as a therapeutic orientation. Contrary to its expansive goals of the nineteenth century in which it sought to become *the* medical system, by 1910, homeopathy offered a much narrower alternative to allopathy; a homeopath was no longer a medical revolutionary, but "*one who adds to his knowledge of medicine a special knowledge of homeopathic therapeutics and observes the law of similia*" (Homeopathic Medical Society of the State of New York 1910, 215). This drastic reduction in homeopathic ambitions was evidence of a stark new reality: homeopaths had lost the epistemic contest and were now exiled to the fringes of medicine.[8]

CONCLUSION—DE-DEMOCRATIZING MEDICINE

By the end of World War I, the reforms of allopathic medicine were complete, and the organizational infrastructure and the modern medical epistemology we know today were more or less in place. The epistemic opening offered by the early cholera epidemics had closed. Regulars, with their labs, were now the unquestioned authority on disease. Per Shakespeare's vision, cholera became a disease identified via the laboratory and treated by allopathic means. While the victory's impact on cholera was minimal—1892 was the last time the disease threatened the United States—it had great ramifications for future epidemics and medicine in general. Regulars no longer had to compete with others over the ownership and definition of disease. Allopaths had achieved epistemic closure.

The process by which bacteriological reformers achieved epistemic closure was all the more impressive, given that it was accomplished outside the bounds of the state. By carrying out their reforms through philanthropic funding, bacteriologists replaced contentious debate with allopathic injunctions, sanctioned, not by public fiat, but by private philanthropic resources. In adopting this "strategy of non-dialogue" (Biagioli 1994, 216), bacteriologists removed medical debates from the public sphere and instead allied with a handful of rich philanthropists to carry out their reforms. In other words, tapping into these private resources allowed bacteriological reformers to circumvent the state and public institutions and to avoid the repeated pitfalls that followed their advocacy in these public institutions. Crucial decisions that shaped the future of medicine were made by elites without general public input. Regulars won the debate over disease by ignoring it, or at least containing the debate among friends in the Rockefeller Foundation boardroom. By 1929, the Rockefeller Foundation had contributed $129 million to medical education and research. Adjusting for inflation, this is equivalent to approximately $1.6 billion today. Alternative sects and resistant regulars simply could not compete with these resources. No one reform—education, public health, hospital, licensing—was decisive but when taken together, they made it extremely difficult for alternative sects to get any foothold in the new organizations of modern medicine. As the momentum for the reforms grew, as more and more organizations came under the rule of the laboratory, the state began to tacitly, if not formally, endorse allopathy. But these endorsements came largely after the fact. And when they were granted, legislators willingly "acceded to physicians' contentions that successful practice required freedom from lay control" (Katz 2002, 31). The technical complexity of the laboratory became a legislative rationale for allowing allopaths to continue their consolidation of American medicine without outside interference.

Because the public—as represented by state institutions—had no opportunity to give its input, the process by which allopathy achieved epistemic closure was problematic in a democratic sense. The successful silencing of medical debates and the consolidation of epistemic authority is all the more striking considering that allopathic consolidation of U.S. medicine went against the public will that had been asserted for nearly a century. Beginning in 1830, state legislatures and the American citizenry had resisted all attempts by allopathy to seize complete control over medical decisions and medical knowledge. They demanded recognition as legitimate knowers and assessors of this knowledge. Unwilling to forfeit these rights, they widely supported the democratized epistemologies of alternative medical sects. Ultimately, Rockefeller money and resources allowed regulars to establish a constellation of specific spheres of professional authority *outside* of state influence. With private support, regulars could avoid the *public* democratic institutions that had long given them problems. There was no need to debate the merits of competing medical epistemologies in the public sphere when a handful of elite physicians could convince a handful of elite philanthropists to fund their programs. The new medical epistemology insulated expert knowledge in a way that represented a challenge to the epistemology of democracy—an epistemology long shared by state legislatures and alternative medical sects like homeopathy which championed competition, open debate, and public participation in medical knowledge.

The allopathic program of de-democratizing knowledge was, in a sense, built into the very fabric of the epistemology of the lab. The laboratory provided regulars with what Steven Fuller (2002,182) calls a "socially protected space" set apart from the democratic process where they could control medical investigation. Knowledge in the lab is largely closed to public observation or oversight. The laboratory does not tolerate public meddling. This insularity of the laboratory is glaring, especially when contrasted to the "space" of bedside empiricism, the patient's own domestic environment, where patients had more power to dictate treatment and speak their minds.

When removed to hospitals, patients, once treated as a source of valuable information about disease, came to be seen as an obstacle to diagnosis. Their words were devalued for scientific tests, their ailments objectified as technical problems to be solved by science. Their status diminished, patients' agency in shaping their treatment was reduced. It is important to note that this exclusionary model of elite knowledge mirrored almost exactly the arguments made by regulars advocating for a privileged professional recognition throughout the nineteenth century. The difference now, after 1892, was that they had the lab and had found an audience amenable to the lab's promise (i.e., rich philanthropists). Using Rockefeller money to reorganize the institutions of American medicine around the lab, they insulated these institutions from public scrutiny. The more professional and scientific medicine became, the less democratic it became. The laboratory provided regulars' long-desired autonomy. By controlling access to the privileged space upon which all of medicine had been reorganized, allopathic physicians controlled medicine.

In the end, the consolidation of allopathic epistemic authority reflected an inherent tension between expertise and democracy. As Eliot Freidson (1970, 336) argues, "The relation of the expert to the modern society seems in fact to be one of the central problems of our time, for at its heart lie issues of democracy and freedom and of the degree to which ordinary men can shape the character of their own lives. The more decisions are made by experts, the less they can be made by laymen." This tension between democracy and medical reform based on expertise was not lost on reformers. In arguing for a dramatic decrease in the number of medical schools, Flexner understood that if his suggestions were taken up, it would result in a more elite profession as lower-class, female, and black students, like alternative physicians, would be locked out of the new system of medicine. But, he argued, restricting the freedom to practice medicine would lead to a net gain in liberty:

> [The community's] liberty is indeed clipped. As a result, however, more competent doctors being trained under the auspices of the state itself, the public health is improved; the physical well-being of the wage-worker is heightened; and a restriction put upon the liberty, so-called, of a dozen doctors increases the effectual liberty of all other citizens. Has democracy, then, really suffered a set-back? Reorganization along rational lines involves the strengthening, not the weakening, of democratic principle,

> because it tends to provide the conditions upon which well-being and effectual liberty depend. (Flexner 1910, 155)

This was a vision for a different kind of democracy. Insofar as the model organization of knowledge shares a constitutive relationship with the model of the good society (Shapin 2008), the epistemology of the laboratory and its accompanying organizational structure offered a new vision of the good society, one that represented a shift from the raucous but open egalitarian democracy of the Jacksonian period to an expert-guided rational polity of the Progressive Era. The participation of the public in medicine (as in politics) would be restricted. Its new role was more limited. In this way, the consolidation of allopathic authority under the epistemology of the lab shaped, and was shaped by, the changing understanding of American democracy and the role of the citizenry in that democracy. Rather than advocating participation, the new vision of democracy, and medicine, coached deference to expertise. Gone were the days when the citizen/patient would be invited to take a more active role in the execution of policy and health. Allopathy had captured cholera—and epistemic authority—in the lab.

CONCLUSION

Medicine after the Time of Cholera

IN 1921, A NEW HERO OF MEDICINE CAPTURED THE IMAGINAtion of the American public—a hero that would have been unrecognizable in the nineteenth century. Although many doctors vied for the role, it was a fictional character that came to personify the ethos of the new epistemology of the laboratory. Martin Arrowsmith, the protagonist in Sinclair Lewis's Pulitzer Prize–winning novel, *Arrowsmith*, embodied the ideals of the burgeoning medical science, his own intellectual growth mirroring that of medicine generally (Fangerau 2006). Indeed, Lewis consulted with Paul de Kruif, a former microbiologist at the Rockefeller Institute, in writing his novel, the intent of which was to paint a satirical picture of turn-of-the-century medicine, while championing the laboratory as the escape from the backward and unenlightened practices of traditional medicine.[1] Imbued with the idealism of the new scientific era, Lewis produced certainly the most popular, albeit saccharine,[2] account of the intertwined processes of epistemic change and professionalization, with the corresponding shift in the identity of regular physicians.

The novel recounts Arrowsmith's peregrinations through the turbulent waters of American medicine. His journey mirrors that of the profession generally over the course of the late nineteenth and early twentieth centuries. Arrowsmith was born in 1883, the year Koch announced the discovery of comma bacillus. His life unfolds in the context of great changes to American medicine, thus offering "the recapitulation in one man's life of the development of medicine in the United States" (Rosenberg 1963, 450). A Midwestern doctor motivated by noble ideals, Arrowsmith struggles with his calling; he wants to help his fellow Americans, but is unsure how best to go about it. Throughout his career, Arrowsmith encounters various potential role models, who serve as signifiers of particular eras and epistemologies in U.S. medical history. Doc Vickerson, a country doctor whom Arrowsmith assists during his youth, is the epitome of the gentleman doctor outlined in the first chapter of this book. Though earnest, Vickerson's practice is limited by his provincialism and his ignorance; his "physician's library" contains only three volumes: "Gray's Anatomy and [the] Bible and Shakespeare" (Lewis 2008, 4). Admiring Vickerson's dedication but recognizing the limits of his hidebound know-how, Arrowsmith departs for medical school, enticed by the new advances in medical sciences. However, there he meets faculty and students who lack curiosity and are "simply learning a trade" (Lewis 2008, 24). Representing clinicians for whom medicine was an art, Arrowsmith finds his medical peers ultimately undignified, both too commercial in their goals and too dismissive of the new laboratory sciences. Upon graduation, Arrowsmith enters the world of public health, but finds the boosterism and the politicking beneath him. He is troubled by the way in which politics muddles, and often undermines, his quest for improving the health of the community. Disillusioned, Arrowsmith is once again set adrift.

Eventually, Arrowsmith finds his calling in research, completing his professional journey, like medicine generally, at the laboratory. Here his role model is Dr. Max Gottlieb, a German bacteriologist so devoted to science that he neglects all social niceties.[3] Lauding the "necessity of technique" (Lewis 2008, 33) and the "beautiful dullness of long labors" (Lewis 2008, 57), Gottlieb is the consummate laboratory scientist. It is in the voice of Gottlieb's thick German accent that Lewis (2008, 278–279) provides what is perhaps the most romantic statement of the scientific ethos underlying the epistemology of the laboratory:

> To be a scientist—it is not just a different job, so that a man should choose between being a scientist and being an explorer or a bond-salesman or a physician or a king or a farmer. It is a tangle of ver-y obscure emotions, like mysticism, or wanting to write poetry; it makes its victims all different from the good normal man. The normal man, he does not care much what he does except that he should eat and sleep and make love. But the scientist is intensely religious—he is so religious that he will not accept quarter-truths, because they are an insult to his faith. . . . He wants that everything should be subject to inexorable laws. . . . He speaks no meaner of the ridiculous faith-healers and chiropractors than he does of the doctors that want to snatch our science before it is tested and rush around hoping they heal people, and spoiling all the clues with their footsteps; and worse than the men like hogs, worse than the imbeciles who have not

> even heard of science, he hates pseudo-scientists, guess-scientists—like these psycho-analysts; and worse than those comic dream-scientists he hates the men that are allowed in a clean kingdom like biology but know only one textbook and how to lecture to nincompoops all so popular! He is the only real revolutionary, the authentic scientist, because he alone knows how liddle [*sic*] he knows. . . . But once again always remember that not all the men who work at science are scientists. So few! The rest—secretaries, press-agents, camp-followers! To be a scientist is like being a Goethe: it is born in you.

It is this identity of the intrepid scientist, unique and set apart, motivated only by an unwavering passion for truth and guided by excellent technique that Arrowsmith heroically adopts. After spending some time in the intensely competitive McGurk Institute—a fictionalized Rockefeller Institute—Arrowsmith retreats to Vermont and admirably spends the remainder of his life in solitary bliss, tinkering in his own personal laboratory.

Arrowsmith offers the ideal representation of the new medical science. The dramatic transformation of medical epistemology that the novel recounts through the figure of Arrowsmith altered the identity of regular physicians. Those trained in Flexner-legitimated medical schools and educated in biomedical sciences focused their aspirations on acquiring scientific expertise, the defining feature of what it meant to be a physician. Reconstituted as new "epistemic subjects" (Knorr-Cetina 1999), regular physicians embraced the identity of the scientist, if not always in practice, certainly in ideals. The model of the doctor was no longer that of the learned man of the community who intimately knew not only his patients' ails but also their character. Nor was it the bedside observer rejecting speculation for the careful, dutiful observation of patient symptoms. And it certainly was not the homeopath with his infinitesimal doses. The one-size-fits-all model of medical training promoted by the Flexner Report demanded that all medical students be indoctrinated into the laboratory in order to become doctors. Bedside manner was de-emphasized for scientific acumen that was attuned to assessing technical information. The wily clinician gave way to the sober, detached scientist, who knew his place by locating himself in the universe of medical science.

Regulars were not the only ones to undergo a change in identity. A similar narrowing occurred for cholera, as its long meandering journey through various identities came to a decisive resolution. Over the course of the nineteenth century, cholera was transformed from a truly heterogeneous thing into a single germ. Once understood as a supernatural scourge, then a cluster of particular manifest symptoms, and later an aggregation of environmental degradations, cholera was radically altered by bacteriology. In spite of the ambiguities of the bacteriological research on cholera, it was reduced to an S-shaped microbe, observable only through the microscope. The presence or absence of this particular microbe determined the presence or absence of the disease. Environmental factors, so crucial in determining the spread of the disease, were de-emphasized (Dubos 1987). Cholera, long a vexing puzzle that "mocked the calculations of man," was apprehended in the laboratory, tamed in petri dishes, and exposed by the microscope. Doctors now knew, or believed they knew, their microscopic enemy and attacked it as such.

Thus, epistemic closure brought a narrowing in the meanings of what was a disease and who was a doctor. Both cholera and allopathy had come a long way since 1832. Arrowsmith, like cholera, had been captured by the laboratory.

This book explains how this epistemic closure was achieved—how regulars came to embrace the epistemology of the laboratory and reconstructed U.S. medicine accordingly—despite resistance from alternative medical movements and government institutions reluctant to grant professional privileges lest they undermine core democratic values. The cholera epidemics, beginning in 1832, created a medical crisis, as allopathic physicians labored impotently to quell people's fears of the foreign disease. This crisis was seized by alternative medical movements, first Thomsonism and then (and more persistently) homeopathy, to reverse the professional privileges that regulars had accumulated in the early part of the nineteenth century. With the public reeling from cholera and wary of regulars' heroic treatments, these movements mounted an epistemological challenge to allopathy. Whether through appeals to make every person his own physician or by demonstrating claims through statistical rhetoric, they proffered more democratic epistemologies that resonated in state legislatures undergoing their own democratic transformation. Alternative medical movements won the near universal appeal of licensing laws; the medical market was deregulated; and an epistemic contest was born.

Facing such a challenge, regulars were forced to articulate a new epistemological vision to replace the discredited rationalism of the past. In response to the successes of alternative movements, reformers within regular medicine sought to capture some democratic bona fides by embracing an epistemology of radical empiricism, derived from the Paris School of medicine. Defined in opposition to rationalism, radical empiricism derided all theorizing, hypothesizing, and generalizing. This left allopaths with a fragmented knowledge base of competing claims and no mechanisms to adjudicate between them; their knowledge on cholera devolved into a series of disconnected empirical observations. To address this problem of adjudication, regulars established the American Medical Association, instituting a no consultation clause that defined homeopathic knowledge as outside of the universe of legitimate knowledge. Regulars tried—and generally failed—to use the organizational leverage of the AMA to prevent the homeopaths from translating government recognition into tangible resources. Still, in the AMA, allopaths formed a united front against "quacks" by constructing an organizational infrastructure insulated from their influence.

With the establishment of boards of health in 1866, a coalition of doctors and sanitary reformers began to rein in cholera's worst excesses. United around a common understanding of cholera as filth, this eclectic group of lay and medical reformers mobilized to prevent cholera by cleaning up the environment and instituting broad sanitary reforms. This was accomplished to great effect. Deploying new technologies like dot maps, the Metropolitan Board of Health in New York City—the board which became the model for other municipalities—was widely credited with stifling the 1866 cholera epidemic. Regulars sought to harness the popularity of the new boards to advance their professional agenda by seizing control of their management. However, with public health expertise framed in terms of its apolitical character and disease framed broadly as filth, this strategy was seen by legislatures as crass politics and contradictory to the ecumenical spirit of the public health movement. It failed; homeopaths and other sanitarians won inclusion on the boards; and public health, much to the chagrin of the AMA, remained an eclectic movement.

In 1883 Robert Koch announced his discovery of the cholera microbe, once again redefining the identity of the disease. The event of Koch's announcement was folded into, and interpreted through, the ongoing epistemic contest in the United States. Initially, both homeopaths and regulars attempted to configure Koch's research as a discovery that legitimated their system of medicine. Through discovery narratives, both sought to embed Koch's research within the history of their respective sects, while simultaneously promoting its promise by depicting it as a pivotal break from the past. Regular physicians proved more successful along these lines. Their narrative provided the intellectual rationale for the construction of a network linked to German science. Bacteriological reformers, most notably William Welch, established enduring connections to Koch's laboratory and, in turn, created a space within regular medicine in which an epistemology of the laboratory could be incubated. Once the discovery became associated with allopathy, homeopaths adopted an oppositional stance against bacteriology that would ultimately doom their popularity as the germ theory gained public acclaim.

With their ties to the German laboratory, bacteriologists embarked upon an ambitious program of medical reform. They sought to remake the entire medical system around an insular, expert-centric epistemology, located in the laboratory, manifest in the new science of bacteriology, and exemplified by the Johns Hopkins model of medical education. Frustrated by the politicized response to the 1892 cholera epidemic in which the dictates of the bacteriological lab were repeatedly ignored by officials, reformers reevaluated their professionalization strategy. Continuously defeated in government institutions on democratic grounds, the AMA circumvented these institutions by aligning itself with large private philanthropies in order to achieve professionalization. Its epistemology of the lab resonated with Rockefeller, who was also integrating the laboratory into the manufacturing processes of Standard Oil. By convincing philanthropies, especially the Rockefeller Foundation, to fund its program of scientific medicine, the AMA evaded state legislatures to construct an organizational infrastructure orbiting around the lab—an infrastructure under AMA control and purified of homeopathic influence. Leveraging their new organizational power, regulars then won special professional privileges from the state *after* these organizations were in place. The nearly century-long dispute over medical epistemology came to an end, as regulars achieved epistemic closure, and in the process, created the U.S. medical professional, unique to Western developed nations in its authority, suspicion of government intervention, and extreme embrace of a scientific-technological approach to medicine.

PROFESSIONALIZATION AND EPISTEMIC CONTESTS

This particular course of U.S. medical professionalization might lead one to mistakenly believe that in the end medical debates did not matter; the AMA in effect purchased its power, using the tremendous economic resources at its disposal. But such a belief truncates history, neglecting to acknowledge the history of defeats that led the AMA to this particular professional strategy. It erases the important contribution of alternative medical movements in creating an intellectual and professional crisis that begged resolution. And it fails to offer an adequate explanation as to how the odd alliance between the AMA and the Rockefeller Foundation came to be. If we want to understand the history of a particular institution—in this case the American medical profession—we must be willing to embrace history in all its messiness and avoid reducing the complexities of the past to simple, monocausal explanations. The meandering course of U.S. medical professionalization was determined by the contours of an epistemic contest. Though waged through specific issues, like cholera, the competition between regulars and alternative medical sects revolved around basic debates over what constituted legitimate medical knowledge. Therefore, to understand the course of the American medical profession—and its unusual outcome—it is essential to recognize this basic epistemic dimension.

This book explores the professionalization of American medicine through the lens of epistemological change, so as to account for the emergence of the laboratory, the nature of epistemic closure in medicine in the United States, and the organizational infrastructure that such closure yielded. The power of professions is rooted in their claim to expert knowledge (Abbott 1988), which justifies a "market shelter" over certain specialized areas of work (Freidson 2001). Professional legitimacy demands a solid intellectual foundation. Given the centrality of knowledge in the establishment, justification, and logic of professions, the analysis of their emergence, and indeed their persistence, must maintain a focus on developments in knowledge. This is not to discount the relevance of other factors (e.g., organizational development, monopolistic work practices, cultural practices, class dynamics, political struggle, etc.) commonly explored in the professions literature. Indeed, these factors arise repeatedly in the analysis in the book. Rather, it is to recognize that, in this case, these factors were fundamentally determined by the underlying issues of knowledge debate and claims-making; to understand the professions and their concomitant organizational structures, political alliances, and labor practices, we must attend to issues of knowledge first and foremost. Because knowledge is the "currency of competition" (Abbott 1988, 102), an adequate sociology of work—and by extension an adequate sociology of professions—must also be "a sociology of knowledge" (Freidson 2001, 27).

The analytical implication of this is clear: the sociology of knowledge must be brought to bear on the study of professions. Unfortunately, the dialogue between these two subfields has been lacking. Given that the professions are the primary organizational form by which developed countries organize expert knowledge, the absence of such a dialogue is both puzzling and inexcusable. Perhaps it stems from the sociology of knowledge's focus on *scientific* knowledge in the past few decades, one more attuned to issues of knowledge production that is somewhat divorced from practical work activities. Perhaps parsing the activities of scientists in the lab has distracted the research agenda away from more clinical and practical forms of knowing. Whatever the rationale for this lack of dialogue—whatever its practical disciplinary justifications—this collective research decision, conscious or not, is problematic. This book opens an analytical space for such a dialogue.

While professionalization is at root always about knowledge, it is not always about epistemology. To be clear: professional disputes—or "jurisdiction disputes" (Abbott 1988)—do not necessarily involve epistemological issues. They can revolve around ownership of a particular work task, where the issues center not on fundamental epistemological questions, but rather on the control of specific tasks or definitions of specific problems (Abbott 1988). For example, disputes between neurologists and psychiatrists in the late 1800s did not focus on questions of psychiatric knowledge—where there was quite an overlap between the two groups—but rather on who controlled the treatment over "nervous" disorders and where the locus (e.g., asylums) of such treatment should be (Rosenberg 1995). Likewise, epistemic contests need not involve the professions. For example, disputes between creationists and evolutionists do not revolve around the issue of professional authority, but rather on things like the composition of the school curriculum. Epistemic contests and professional disputes can overlap, but they are distinct.

This distinction becomes evident when comparing the process of medical professionalization in the United States to that of other national contexts. While claims to American exceptionalism are often rife with exaggerations and normative assertions, the American medical profession has in fact departed from its peers in other Western democracies. The most obvious—and politically contentious difference—is its approach to health insurance, where "the United States stands out for the virulence of its political battles over health care" (Starr 2011, 1). Prior to the passage of the Patient Protection and Affordable Care Act (PPACA) in 2010, the United States was the only advanced country to lack a government-mandated system for universal health care. And even the mechanism by which PPACA seeks to achieve universal coverage is unique in that it does so largely through the private sector. The country's approach to medical insurance is in great part a reflection of the profession's adamant opposition to government intervention in medicine.

The absence of some sort of government-backed mechanism to achieve universal health care coverage is not the only way in which the U.S. medical system diverges from other developed countries. The American medical profession is also exceptional in its focus on technologically intensive medical services, its commitment to specialization, and historically, its embrace of the germ theory. To an extent unusual in the early twentieth century, American medicine was built around the promises of technological fixes, a tendency that has been exacerbated over the years by the structure of U.S. medical economics, which favors expensive technological interventions (see Birenbaum 2002, Clarke et al. 2003).

I have labored to show that the U.S. medical profession's embrace of the epistemology of the laboratory and its suspicion of government institutions—the source of much of its exceptionalism—resulted from the specific trajectory of the epistemic contest. This argument raises a critical question when teasing out the relationship between epistemological change and professionalization: if all Western countries eventually embraced the laboratory sciences, why did this epistemic shift yield such different organizational/professional outcomes in the United States? Doesn't it undermine the significance of the relationship between professionalization and epistemic change if similar epistemic outcomes lead to different professional ones?

The answer, I argue, is not really. While it is beyond the scope of this book to offer a comprehensive comparative account of professionalization in different contexts, a few comments are in order. First, as the status of the United States grew in the world, so too did its medical influence over other countries. It became a major exporter of the epistemology of the laboratory, and its doctors global emissaries for the lab. The worldwide adoption of this epistemology resulted in part from the U.S. epistemic contest, which created a profession whose very identity hinged on the laboratory and offered a vision that was exported to other countries via philanthropic projects. Second, the epistemological similarity between different countries leads many to gloss over real differences in the *extent* to which physicians in other nations embraced the epistemology of the laboratory. My argument vis-à-vis the exceptionalism of the U.S. case is a matter of degree, not kind. All Western countries eventually adopted the epistemology of the laboratory, but none with as much fervor as the United States. The germ theory became "gospel" in the United States (Tomes 1998) to an extent unrealized elsewhere. This degree of commitment reflected the particular nature of the epistemic contest in the United States. The identity of the medical profession in the United States came to be built on the laboratory; this was simply not the case in other countries (e.g., England), where professional political dynamics differed. While other countries demonstrated a greater emphasis on public health, general medicine, and prevention, the U.S. medical profession poured its energies and optimism into scientific medicine, as the laboratory justified the profession's power. Therefore, while there was overlap between national contexts (partially due to the fact that they were not isolated from each other), the trajectory of professionalization in each country contained its own idiosyncrasies that subsequently became inscribed in their medical systems.

Ultimately, the relationship between epistemic contests and professionalization is a historically contingent one, not a necessary one. The exceptionality of the U.S. medical system emerged out of (and can only be understood as resulting from) the unique history of its epistemic contest. This epistemic contest—with a strong democratic culture resistant to professions, active challenges by alternative medical movements, and eventual consolidation through the unique system of U.S. philanthropy—meant that the epistemology of the lab became linked to a suspicion of government intervention. But this was a specific outcome, contingent on a specific confluence of factors.

The professionalization of medicine in other countries did not involve epistemic challenges of the same nature or intensity. Context matters. The way in which different constellations of actors intersected in the political, social, and cultural systems of various countries affected the degree to which bacteriology mattered for the organization of medicine.[4] For example, in Great Britain, the medical profession evolved along class lines (Shortt 1983). Elite physicians shared a background with government elites and always had their support. Although initially questioned for geopolitical reasons, the germ theory was folded into the existing professional hierarchy that had long-standing support from the state. In France, the germ theory was embraced first by a strong central government, which, given its reach, brought reluctant doctors along (Latour 1988). Indeed, the process of medical professionalization in France was carried out *through* the state, occurring much earlier than in the United States (see Ellis 1990; Geison 1984; Goldstein 1990; Weisz 1978). In other words, the adoption of bacteriology and the epistemology of the lab in England and France did not have the professionally transformative character that it had in the United States. In fact, of the developed Western European countries, only Germany had a medical profession that demonstrated a similar wariness toward working with the state. This reluctance, however, was resolved during the nationalistic run-up to World War I (Kater 1985). In the end, although rigorous comparative research on the professionalization of medicine in different countries is needed to tease out the nuances of these differences and similarities, the history presented in this book demonstrates the unusual extent to which the professional politics of U.S. medicine were animated by basic epistemological issues. In no other country did medical professionalization involve either the same epistemic dimensions present in the United States and/or the same hostility from state institutions.

When professionalization and epistemic contests do overlap, as they did in the case of nineteenth-century medicine, we ignore epistemology at our peril. In these cases, actors must negotiate a confusing hodgepodge of competing knowledge claims. They are forced to deal with epistemological issues that, though ever present, are normally taken for granted. To navigate uncertainty in knowledge, we have institutionalized epistemic standards that supply a structure or framework by which actors can discriminate good "truth" from false "belief." These ever-present criteria for assessing beliefs inform and determine the manner in which individuals make sense of reality. In most cases, for most people, these criteria remain unarticulated. An individual does not need to explicitly know justificatory arguments to employ them in the pursuit of knowledge (BonJour 1978). Rather standards are institutionalized in the social practices of knowing.

The concept of the epistemic contest is intended to shed light on those moments of crisis when epistemological standards break down and become contested. It seeks to sort out how certain ways of knowing ("habits of reasoning" [Peirce 1955,123]) become socially established, how actors wage struggles over epistemology, how epistemic closure is realized, and how emergent epistemologies become institutionalized in organizations. Epistemic contests are a particular type of knowledge dispute, in which actors, advocating competing understandings of reality and the nature of knowledge, struggle in various realms to achieve validation to their approach of knowing. They involve questions such as what constitutes a fact; by what standards can true knowledge be distinguished from false belief; what are the conditions by which claims could be said to be justified; what is the relationship between the observer and the external world; and who can be considered a legitimate knower. Thus, rather than debating the merits of particular knowledge claims vis-à-vis a system of agreed-upon standards—as most knowledge debates do—epistemic contests engage with the standards themselves.

Put differently, rather than playing in a game with established rules, those involved in epistemic contests are fighting over the rules themselves. In this sense, epistemic contests are more encompassing, more fraught, and more open-ended than typical knowledge disputes, like credibility contests (Gieryn 1999). They are not just about drawing cultural boundaries between science and nonscience; they are about establishing the parameters for truth and falsity. In turn, the logic of action that epistemic contests compel is different from other forms of knowledge struggles. Actors in epistemic contests must do double work; they must find ways to justify *both* their particular truth claims *and* the epistemic assumptions embedded in those claims. In the case of nineteenth-century medicine, competing medical sects not only had to promote their ideas regarding cholera but also had to fight for the assumptions about the nature of medical knowledge that undergirded their ideas. Rather than fighting for credibility within an established, shared system of epistemic values, medical sects engaged in fundamental debates, whose contours were ill-defined.

The openness of epistemic contests results in a wider array of strategies deployed. One of the striking elements of the history of the epistemic contests in the United States is the extent to which the various medical sects deployed organizational strategies to promote and solidify their epistemological positions. The sociology of science downstream focuses on cultural strategies like boundary work (Gieryn 1983, 1999), performance (Hilgartner 2000), and rhetoric (Gilbert and Mulkay 1984), which are certainly but by no means sufficient in waging epistemic contests. Epistemic contests are not waged by cultural means alone. Indeed, because epistemic contests occur in a densely populated organizational terrain, organizational strategies become very important as actors try to harness organizations to legitimate their epistemological systems. One way stakeholders adjudicate knowledge claims in epistemic contests is *through* organizations, which stamp some knowledge as legitimate and others as outside of the realm of consideration. This is most evident in chapter 2, which discusses how the establishment of the AMA and the institutionalization of a no consultation clause represented an attempt to address the problem of adjudication via organizational practices in an environment bereft of epistemic standards. Organizational formation, in this case, represented an epistemological strategy. The upshot of this analysis, therefore, is that the nature and trajectory of epistemic contests can only be understood as emerging from the interaction between cultural and organizational practices.

These strategic interactions are always embedded in particular contexts, or arenas, that affect their outcomes and the trajectory of epistemic contests more generally. In considering different epistemologies, it is important to avoid the tendency to reify them as free-floating intellectual systems. Rather, they are arguments that arise in particular institutional settings, settings that influence the degree to which they are embraced or rejected. For example, the context of the state legislatures affected the epistemic contest over medical knowledge in that legislatures were more sympathetic to the democratized epistemologies of alternative medical sects. In a sense, the entire epistemic contest over nineteenth-century medicine can be read (without doing much damage to the nuance of the analysis) as a failed struggle for regulars to achieve a privileged position in government institutions that were committed to democracy in knowing and therefore suspicious of regulars' professional aspirations.

The metaphor of a contest is therefore intentional. Epistemological systems do not originate fully formed. Rather they arise through competition and the strategic interaction among actors. This book offers an account of the development of medical knowledge through conflict. Poked and prodded by alternative medical movements, regulars were forced to provide an epistemological justification for their authority. Recognizing the problems with rationalism, reformers adopted first radical empiricism, and later an epistemology based on the laboratory. Alternative medical movements responded to these reforms and the strategies that followed from them. Epistemic change followed the give-and-take strategic dance between competing actors. The concept of the epistemic contest is thus situated in a broader call for sociology to take seriously the analysis of embedded strategic action (Jasper 2006).

RETHINKING AMERICAN MEDICAL PROFESSIONALIZATION

The benefit of examining the epistemological foundation of the professionalization of medicine in the United States is that it leads to empirical findings that help flesh out our understanding of the U.S. medical system and the precarious place it finds itself in today. The proof of a framework's usefulness is in its explanatory pudding. What does looking at this case of professionalization through an epistemological lens tells us empirically that we would not have seen otherwise?

First, and most broadly, the analysis in this book specifies the mechanisms that undergird the macro-cultural account of U.S. medical professionalization. It eschews the analytical laziness of labeling an era and calling it an explanation. By focusing on practice, it allows for the identification and explanation of the human action that comprises such macro-cultural shifts. Certainly the cultural changes that accompanied the Jacksonian era played a role in creating an environment conducive to the epistemic challenge mustered by alternative medical movements. And undoubtedly the burgeoning acceptance of expert knowledge during the Progressive period was favorable to the reform of American medicine through the laboratory, exemplified in the Flexner Report. But these general shifts are not sufficient enough to tease out the nature of professionalization, or even to understand its course. Macro-cultural arguments may provide some basic contours of the narrative, but their lack of specificity leaves much of the story untold. By moving to a meso-level analysis, this book shows how large-scale cultural shifts operate through specific practices in specific settings. Rather than making a broad appeal to the culture of the Jacksonian period, it shows the ways in which the types of epistemological visions proffered by homeopaths and Thomsonians resonated in certain government institutions. Culture, in turn, is no longer conceived as something external, hovering above social action that sets the context, but as something produced and reproduced in practice.

Second, this book stresses the essential role of alternative medical movements in the development of medical knowledge. Rather than mere curiosities or, worse, repositories of errors, alternative medical movements emerge from this analysis as crucial. In addition to any specific intellectual influences they had (e.g., convincing regulars to discard heroic therapies like bloodletting), by forcing regulars to legitimate their professional claims in epistemological terms, these movements drove changes in medical knowledge. In other words, because epistemological positions arose from contentious struggle, the influence of alternative medical movements was manifest in the outcome of the epistemic contest. The crucial role played by alternative medical movements has been woefully underestimated in the dominant histories; alternative medical movements drove developments in medical knowledge through their dynamic, magnetic relationship with regulars. This case, then, points to the potential of social movements to play a more substantial, if more subtle, role in influencing knowledge than is typically recognized. Research on "scientific and intellectual movements" reveals the important influence of movements in scientific-knowledge-producing institutions, like research universities and scientific disciplines (Frickel 2004; Frickel and Gross 2005). However, because it largely restricts its analyses to autonomous scientific fields in which activists are forced to conform to scientific epistemic standards so as to portray themselves as credible knowers (see Epstein 1996), this body of research overlooks the potential influence of social movements in fundamental epistemological debates. In less autonomous fields, like medicine, or in fields experiencing epistemological flux, social movements like homeopathy can engage in "knowledge advocacy" on fundamental epistemological grounds (Whooley 2008), offering "epistemic challenges" as they are less pressed to debate the issues on the field's own terms (Hess 2004). In other words, the analytical frame here allows for a wider recognition of the "cognitive praxis" (Eyerman and Jamison 1991) of social movements, one that penetrates to the level of epistemology.

Third, this book clarifies the nature of the alliance between elite philanthropies and medicine by revealing the epistemological affinity between the bacteriological laboratory and Standard Oil's incorporation of chemistry laboratories in its industrial operations to gain a competitive edge. Typically, the alliance is reduced to simple class terms. For example, according to the account offered by E. Richard Brown (1979), Rockefeller adopted medical science as his philanthropic cause both to improve his public image in the wake of damaging labor disputes and because the laboratory portrayed problems as technical issues amenable to expert intervention, rather than general social problems of inequality. These dimensions undoubtedly were at play, but they do not explain why Rockefeller chose medicine specifically, and they overstate the degree to which the alliance was forged on the basis of crass class politics. By illuminating corporate scientific practices, long neglected in the sociology of science (Penders et al. 2009; Shapin

2008), this analysis transforms Rockefeller from a mere cunning capitalist into a cunning capitalist who pioneered the integration of laboratory sciences in industrial practice. His efforts to integrate the laboratory mirrored the efforts of bacteriological reformers, with whom Rockefeller shared an epistemic affinity. And it was upon this affinity that a mutual alliance between the AMA and private interests was built.

The final historical correction that this book achieves relates to this last point. Regular reformers were able to translate this affinity for the laboratory into an alliance that enabled them to circumvent state legislatures to acquire the sufficient funds for their professionalization project. The decision to build the laboratory with *private* resources allowed regulars to avoid the *public* institutions that had rejected their professionalizing impulses for decades. Indeed, the desire to circumvent the government institutions was evident as early as the 1870s, when regulars lamented, "Legislative enactments in the various States of this Union clearly show that no reliance can be placed on either the uniformity or permanency of any laws now relating to the practice of medicine" (Medical Society of the State of New York 1870, 39). The "melancholy illustration" (Hutchinson 1867, 56) of continual failures in its legislative agenda led the AMA to seek an alternative route to professionalization, one that skirted government institutions. In its very establishment, the modern U.S. medical profession expressed hostility toward government intervention and wariness of working through state legislatures. This points to an earlier emergence of the AMA's antigovernment sentiment—and its corresponding embrace of private interests—than is typically acknowledged in historical accounts of the U.S. profession. Most histories of the U.S. debates over health care trace the AMA's wariness of state intervention to the World War I period (see Numbers 1978); their histories start in 1915 (see Starr 2011). This finding is more than the mere dating of a phenomenon. I demonstrate how the antigovernment sentiment was present in the very founding of the profession and as such, institutionalized in its professional culture. Given that regular physicians achieved professionalization *in spite* of the state, the hostility toward government intervention was inscribed in the very DNA of the profession. The strategy of achieving their ends through private means became a tried-and-true one postprofessionalization, one with a history of success. Thus, the exceptionalism of the U.S. medical system in its rejection of government intervention, on display most glaringly in the AMA's persistent campaign against government-run health insurance, has its roots in the manner in which the epistemic contest over medicine played out in the nineteenth century.

To be clear, I am not suggesting that the entirety of the muddled and vitriolic history of health insurance reforms in the United States is solely the result of the epistemic contest; I am not offering a mono-causal account of the decidedly peculiar U.S. medical system. Health care politics in the United States are exceedingly complex, and to attribute the entirety of this complexity to a single factor would be audaciously reductionistic. A number of factors (e.g., a weak labor movement, an impotent socialist political party, the federal system, resistance from insurance companies, a cultural of individualism, etc.) contributed to the long-standing resistance toward government-led health insurance. Indeed, the AMA, with its suspicion of government intervention and its early alliance with private industry, played an important role in shaping this system, but it was only one interest among many (albeit an important one). Furthermore, the U.S. health care system has experienced a number of dramatic organizational changes (Scott et al. 2000), character redefinitions (Light and Levine 1988), and market shifts (Timmermans and Oh 2010) since World War II.

In recounting the "peculiar" history of health care politics, Paul Starr (2011) outlines a two-stage model of the politics of U.S. health care. During the first half of the twentieth century, the politics were driven largely by interests groups, especially the AMA, which rejected government incursion into medicine, often framing such oppositions in terms of anticommunist ideology. This early era established the "script" for health care debates for the rest of the century, as evidenced by the repeating tropes of "socialized medicine." During the second period, dated roughly from the 1950s, politics were constrained by what Starr (2011) terms a "policy trap." In other words, the debates of the last half century have been constrained by the original mishmash of policies from the first half. Rather than approaching the issue broadly, the terms of the debate narrowed significantly, making broad reform efforts increasingly impossible. Accepting Starr's two-stage model of the history of U.S. health care reform efforts, the legacy of the epistemic contest was felt most strongly during the first period, when the AMA established itself as a staunch opponent to government-led health insurance.[5] It was during this time that the AMA's wariness of the state—the legacy of the epistemic contest—dominated the rationale of the AMA. Only a few decades removed from the wide open medical market of the nineteenth century, allopathic practitioners were unwilling to cede any of their hard-won professional authority, especially to government entities that had repeatedly denied the legitimacy of this authority on democratic grounds. Of course the politics of the new era were different; the more sedate politics of the Progressive Era had supplanted the woolly democratic experiments of the Jacksonian period. But the wounds remained for regulars, and having achieved epistemic closure *in spite* of the state, the AMA was unwilling to let it back in and fought tooth and nail against such a fate.

Insofar as the AMA was one of the most powerful stakeholders that successfully defeated early reforms efforts (Quadagno 2005), which later set the script for health care debates, and insofar as this oppositional stance was born of the epistemic contest, we can see the long shadow cast by nineteenth-century epistemological debates inscribed in the modern U.S. health care system. By attending to issues of epistemology and investigating the professionalization of U.S. medicine as a case of an epistemic contest, this book offers a more exacting account of the history of the American medical profession—one that tells us as much about the present as the past.

DEMOCRACY, PROFESSIONALIZATION, AND EPISTEMIC CLOSURE

By the 1920s, regular physicians, through the AMA, had achieved epistemic closure. This is not to suggest that alternative medical movements ceased to exist. Indeed, alternative approaches to medicine persevered through the twentieth century (Whorton 2002; Young 1967) and have made a comeback in the past two decades (Eisenberg et al. 1998), especially with the contemporary attempts to incorporate Complementary and Alternative Medicine (CAM) into mainstream medicine. Nevertheless, despite this dogged perseverance, the endurance of alternative medicine on the margins does not undermine what Magali Larson (1977, 37) calls the "exceptional character of medicine's professional success." Once epistemic closure was achieved, alternative medical movements were relegated to the fringes of medicine. At the center of the new scientific medicine was the laboratory, and by controlling access to the laboratory, regular physicians had reduced the epistemic threats of alternative medical movements to mere nuisances.

On the whole, the laboratory revolution and the profession's extreme embrace of a biomedical model proved quite productive, although it was not without its problems and blind spots. By the mid-twentieth century, the laboratory took medical science to important new heights, improving diagnosis and treatment. It goes without saying that one would much rather be a patient today or even in 1892, than in 1832 or 1866. The most dramatic benefits came from the investment of time and energy into vaccination, which followed directly from the epistemic closure around the laboratory. While a cholera vaccine never became widespread,[6] bacteriologists did discover effective vaccines[7] for diseases like rabies (1885), tuberculosis (1921), yellow fever (1937), polio (1950), and the measles (1963). In terms of treatment, the diphtheria antitoxin, developed by Emil von Behring in the early 1890s, was bacteriology's first real triumph. Though diphtheria was not that prevalent, it became important symbolically for reform. Bacteriologists heralded the antitoxin as justification for laboratory science, and it would remain the major success story for nearly two decades (Hammonds 1999). Salvarsan, a drug for syphilis developed by Paul Ehrlich in 1909, represented another early victory for bacteriology. Still, it was not until the late 1920s and 1930s, with the introduction of antibacterial drugs, like penicillin, and synthetic sulfa drugs, that bacteriology really bore therapeutic fruit—three decades after Koch's announcement of the discovery of the comma bacillus.

More immediate benefits came in the form of antiseptics and diagnostic technologies. The germ theory provided the explanatory framework that justified sterilization techniques for surgery,[8] which dramatically reduced deaths from sepsis, infections, and putrefaction (Starr 1982; Temkin 1977). As for diagnostic practices, laboratory diagnoses were a boon, in that they seemed to provide conclusive evidence of the presence or absence of disease. Once again this benefit was not immediate as it took awhile to catch on given the technical difficulties and an initial lack of standards in bacteriology (Gossel 1992). Nevertheless, these diagnostic tools solved one of the more persistent issues regarding epidemics—the frequent and destructive early debates over whether a given epidemic disease was present in a locale.

Indeed, wielding the bacteriological model, modern medicine did conquer many infectious diseases, but this was accomplished long after the resolution of the epistemic contest, as these innovations did not become commercially available and widely adopted until after World War II. Between allopathy's embrace of the laboratory and its therapeutic fruits lay nearly fifty years of sparse accomplishments in which the search for what Paul Ehrlich called "magic bullets" ran largely on the fumes of promise. Still, even if we acknowledge the fits and starts of bacteriology—which the truth-wins-out narratives fail to do—the record is impressive. Despite the delay, bacteriology achieved important medical advances. The early promise of bacteriology, while perhaps naïve in its optimistic vision of magic bullets right around the corner, was not misguided.

Nevertheless, while experimental laboratory science led to important medical advances, it also created some real blind spots. In delimiting acceptable knowledge, an epistemology inevitably excludes and/or ignores knowledge that does not fit its standards. The laboratory defines disease in a limited way, reducing it to the presence or absence of a microbe. To understand the limits of this reductionism, we need look no further than the manner in which cholera has been approached in the century since epistemic closure. Public health measures built on the bacteriological paradigm prove problematic when encountering the disease outside of the laboratory. In their excellent analysis of the epidemic in Venezuela, Charles Briggs and Clara Mantini-Briggs (2003) expose the pitfalls of current public health interventions that have their roots in the bacteriological paradigm. In 1992, cholera arrived in the eastern Delta region of Venezuela, hitting its indigenous communities particularly hard. In response, the World Health Organization (WHO)—an organization financed in part by the Rockefeller Foundation—dispatched an impressive number of medical officials to the Delta. The WHO's strategy was built on an understanding of cholera as a germ. They "imbued *Vibrio cholerae* with quasi-military agency" (Briggs and Mantini-Briggs 2003) and approached the epidemic as a problem of germs and individual patients, not as a collective or environmental issue. Determining that the indigenous could not be sufficiently educated to prevent the disease, these officials focused on treatment, dispensing an incredible amount of antibiotics and, to a lesser extent, oral rehydration therapies. In other words, officials attacked the epidemic through the microbe. Though these efforts served to curb the worst excesses of the epidemic, once the crucial period passed, the medical officials quickly departed, having killed off the bacteria. What wasn't addressed by the WHO is telling. There were no improvements made to the inadequate sanitary infrastructure of the Delta, no long-term funding for staffing of medical clinics, and no policies to address the social and economic inequality that enabled the epidemic in the first place. In targeting the germ, officials ignored the environment, leaving the Delta just as vulnerable to future cholera epidemics as when they arrived. They gave pills and left. Compare this to the successful sanitary efforts in the United States 126 years prior, in which diseases like cholera were addressed through environmental improvements, and the limitations of a bacteriological model for public health becomes starkly evident. The more public health is understood solely as an extension of bacteriology, the more limited it is in addressing the underlying social and structural causes of epidemics (Dubos 1987). The limits of the germ theory along these dimensions represent in part an epistemological failure of imagination (Farmer 2001).

In reaction to this reductionism, current research on cholera has sought to undo the rigidity of the bacteriological paradigm by embracing a more complex understanding of the disease. The laboratory has been "home for cholera for much of the twentieth century" (Hamlin 2009, 236), but researchers are now trying to replace the cholera-as-a-germ model with a much messier, more unstable definition, by examining cholera in situ. The ecological view of the cholera microbe and "biocomplexity" recognizes that cholera interacts with its surrounding environment in fundamental ways and its pathogenicity is an acquired state (Colwell 2002; Colwell and Huq 2001). In other words, rather than extracting cholera from its environment to study it, this new paradigm *re-embeds* it to better grasp its nature. This is a direct challenge to the laboratory. And it complicates the identity of the disease; as historian Christopher Hamlin (2009, 16) notes,

> No longer is it possible to claim with confidence to *know* cholera. With cholera changing and inchoate, the control in our social and historical experiments has vanished. What has seemed most solid about it, its microbe, has turned out (like other microbes) to be a repository for varying bits of rogue DNA, which together express toxicity under certain conditions. While we know vastly more about it, the general entity "cholera" is less fixed than at any time since 1830.

Armed with this new (or perhaps old?) perspective, current researchers "wonder why there has been so little work of what seemed obvious questions" (Hamlin 2009, 230); the answer is that the epistemology of the laboratory prohibited (or at least impeded) these questions from even being asked.

This shift in perspective is dramatic, its implications significant. The parsimonious view of cholera as a microbe, inherited from Koch, is giving way to a much messier picture, one that recognizes "many varieties of *Vibrio cholerae*, gaining or losing pathogenicity as toxin-bearing stretches of DNA move in, or toxin expression is turned up or down in response to environment" (Hamlin 2009, 211) . By reintroducing complexity, the blinders that the epistemology of the laboratory erected are breaking down and its power as a means to understand cholera is being questioned.

Expert Knowledge in Democratic Cultures

The manner in which epistemic closure was achieved by regulars not only had ramifications for the conceptualization and treatment of disease entities; it also raised more fundamental issues of professions and democracy. A recurrent theme throughout the book has been the tension between the epistemic/professional project of allopathic medicine and democratic ideals. Prior to the 1900s, the history of allopathic professional struggles in these government institutions was one of the continuous rejection of regulars' claims to privilege and programs of exclusion. Insofar as the American public voted on these issues, through their representatives, their message was clear—openness in the medical market was a positive ideal. The epistemological logic underlying these outcomes was one that viewed debate, dialogue, and transparency as essential to the development of medical knowledge.

Epistemic closure, however, was achieved outside the type of public institutions that facilitate and ensure democratic oversight. Allopathic physicians, wielding premature claims of certainty, won the epistemic contest by convincing elite philanthropists to back their program outside of state influence. Philanthropic resources allowed them to reorganize the medical system around the laboratory and create an infrastructure under their control, which they leveraged to kill off alternative challengers. They effectively purchased the professional privileges they could not win through argumentation. Making the laboratory central to medical knowledge foreclosed democratic debate and took the discretion over medical matters out of the public sphere.

Grounding professional authority in the laboratory placed medical knowledge firmly under the control of regulars. Although the specific means by which this was achieved were particular to medicine, the desired goal to monopolize knowledge so as to gain a market advantage is the essence of all professionalization efforts. In this way, professionalization conflicts with democratic values. The manner in which communities organize the production of knowledge shares a constitutive relationship with the model of the good society (Shapin 2008). How knowledge is to be achieved—and its concomitant hierarchies—mirrors an understanding of how decisions should be made. In the United States, professions have become the primary means by which the marketplace deals with expert knowledge. Certain actors are provided autonomy from outside influence, a privileged insulation granted on behalf of their perceived expertise and their status as elite knowers (Freidson 2001). But expert formal knowledge is not part of everyday knowledge; it is an elite knowledge that by nature is not democratic (Freidson 1970) and therefore exists awkwardly in democratic cultures. Karl Mannheim (1992, 185) recognized long ago, "Democratic cultures have a deep suspicion of all kinds of 'occult' knowledge cultivated in sects and secret coteries." Insofar as their knowledge is opaque and "occult-like," professions represent a threat to democratic ideals in that they have a unique capacity to discourage, distract from, and limit democratic deliberation (Dzur 2004). They carve out a hierarchy of privileged knowers within an (ideally) flattened democratic culture. Knowledge production becomes the province of the few, undertaken in spaces like the laboratory that are closed to the public. Democratic political governance, on the other hand, posits a very different epistemology, emphasizing openness, transparency, and the participation of a general public. The two visions of knowledge that form the basis of democracy and professionalization are in an important sense discordant.

This dissonance must be resolved in practice. In a sense, the entire epistemic contest over nineteenth-century medicine revolved around the struggle between allopaths' promotion of a privileged hierarchy in medical knowledge and alternative medical sects' promotion of more democratic forms of knowing. After nearly a century of coming out on the losing end of the democracy/profession tension, regulars were able to draw on philanthropic resources to circumvent government institutions, achieving their professional goals *outside* the bounds of the state. Only after they had consolidated organizational power did allopathic physicians and the AMA then turn to the state. They were then able to use their newly acquired organizational leverage to dictate the terms of medical legislation and ensure that the oversight of every facet of medicine would be in their control. The AMA's effective resistance to state interventions established the condition for a highly privatized medical system, exceptional throughout the developed world (Krause 1996; Saks 2003). In the end, allopaths resolved the tension between democracy and professionalization by avoiding it altogether, convincing a handful of elite philanthropists, rather than a democratic populace or their representatives, that the laboratory was the only legitimate way to medical knowledge.

The social order of knowing promoted by bacteriological reformers was not just the reflection of cold professional calculation; it was embedded in the very epistemology of the lab. The laboratory is not a democratic space. It rarely allows outsiders in and even if it does, the barriers to participation are so high that only experts can realistically speak to what goes on there. It restricts participation to the technologically elite, who are both the producers and judges of knowledge claims produced therein. The public record of laboratory practices—journal articles—obscures their messiness and ambiguity (Latour and Woolgar 1986). In terms of the universe of knowers, the laboratory offers a restricted, closed system. Those outside are asked to trust its claims even though they are neither allowed, nor able, to adequately assess them. For these reasons, the laboratory was alluring for allopathic professional reformers. Its exclusionary nature fit squarely in a legacy of exclusionary allopathy epistemologies, evident first in authoritative testimony of rationalism and later imposed, under radical empiricism, by the AMA.[9] The difference was that the scientific aura of the laboratory won important allies like Rockefeller, who provided the requisite resources for professionalization through the lab. The democratic arguments against allopathy had not changed. They were not defeated nor repelled. They could simply be ignored because reformers had tapped into a vast source of private resources.

This history has important ramifications for how we think about the role of professions in democratic societies. Because professions are the most common way that modern societies institutionalize expertise, there is a tendency to take them for granted as the natural outcome of living in an increasingly complex world. What this book shows is that rather than natural outcomes, professions are born from historical processes, central to which is the need to resolve the tension between professional claims and democratic cultural values (at least in democratic societies, that is). There is no ideal way to achieve this resolution; rather as professions research has long argued, we need to attend to the historical particularities out of which these tensions are resolved.

The history recounted in this book thus raises difficult questions. What is the place of expert knowledge in democratic societies? How can we organize knowledge to preserve democratic values while still acknowledging the need for hierarchies in knowing? How can we calculate and justify such a trade-off? How can we ensure equality and patient participation, and prevent epistemic inequality, while maintaining patient protections, sensible regulations, and a functioning system of medical credentialing? These types of questions are not restricted to medical issues, as they permeate the heart of the democratic experiment; they can be read into any debates that involve trade-offs between freedom and equality. They are the type of questions that resist neat, abstract answers, but instead are sorted out in the messy world of local practice, a world rife with struggle, capitulations, compromise, and "satisficing" (March and Simon 1958).

To be clear: this is not to say that professions are antidemocratic in their nature. Rather it is to realize their inherent tensions with democratic ideals, tensions that must be negotiated in practice. For medicine, the awkwardness was dealt with via circumvention. Other professions may resolve it differently. But we must attend to this issue in our analysis if we are to understand the emergence and character of particular professions.

The democracy/profession tension speaks to a broader issue of inequality that stems from epistemic closure. In achieving professionalization, professions are granted authority and control over the deployment of expert knowledge (Abbott 1988; Freidson 1970, 1986, 2001). This privileged epistemic authority is granted at the expense of others. As this book illustrates, epistemologies differ in the degree to which they are inclusive of outsider participation. The epistemology of the laboratory falls on the exclusive end of the inclusivity/exclusivity spectrum. As it became institutionalized as *the* way to achieve medical knowledge, this exclusivity was made manifest in medical organizations and practice. Some of this exclusion was intentional. Regulars wanted to ensure that homeopathy was marginalized. But other aspects of this exclusion were less explicit, perhaps less intentional. Epistemic inequality was overlaid by gender and racial inequalities, resulting in the systematic exclusion of certain groups from the universe of legitimate medical knowers. For example, the Flexner Report, in assessing medical schools on laboratory-based criteria, was particularly hard on medical schools that trained minorities and women, which lacked the resources of research institutions to establish such facilities (Starr 1982, 124). Medicine became more homogenous as the profession became even more dominated by white males.

Reason's "entanglements with social power" (Fricker 2007, 3) is a core insight for any sociological study of epistemology, as are the ethical implications of this insight. Ultimately, the exclusionary nature of some epistemologies and the institutionalization of this exclusion in organizations can create disparities in the allocation of trust and credibility afforded to an individual's or group's testimony (Fricker 2007, 1). Some people are excluded from knowing either because their testimonial appeals are ignored or because the hermeneutic systems they draw upon to make claims are misunderstood or devalued (Fricker 2007). The exclusion of some from participating in knowledge production and assessment, whatever the reason, creates power inequalities between the epistemological haves and have-nots.

This epistemic inequality has negative ramifications not only for those in positions of weakness; taken in the aggregate, it can hurt the community in general if potentially fruitful ways of thinking are closed off. Epistemic closure can result in insularity, creating inward-looking communities of knowledge producers, complete with rigid barriers to entry, which fail to transmit knowledge claims beyond a restricted sphere. The insulation of knowledge in particular epistemic communities "can be thought of as the existence of barriers to communication and experience, barriers that facilitate systemic ignorance and misunderstanding, and the coexistence of otherwise conflicting practices, understandings and expectations" (Reay 2010, 92). Epistemic closure, therefore, is not just an issue of access and participation; it becomes one of communication and knowledge production, as it produces structural incompetencies that constrain potential directions for exploration.

In this sense, allopathy's epistemic closure represented a failure of a kind of intellectual humility. For professional, political, and intellectual reasons, allopathic physicians, through the AMA, sought to absorb the uncertainty in medical knowledge into the promises of the laboratory. In the words of William James (1909, 9), allopathic reformers sought "to carve out order by leaving the disorderly parts out." This endeavor had two components. Intellectually and epistemologically, the laboratory offered a controlled environment bereft of the messy uncertainty of nature. It literally carved out a purified space so as to achieve scientific certainty. By forcing all medical knowledge to conform to a single epistemological system, allopathic reformers vanquished dissenting voices. In doing so, they intentionally stifled the type of debate, dialogue, and interaction that may have borne intellectual and therapeutic fruit. This is not to dismiss the benefits achieved by the laboratory. But it is to worry about the lost opportunities—opportunities that a system more in accord with democratic values might have realized.

TOWARD A SOCIOLOGY OF EPISTEMOLOGIES

The main theoretical intervention of this book is to get sociologists thinking about, and engaging with, epistemology. Key to this is the observation that epistemological issues and debates are not confined to the rarefied air of academic philosophy, but rather are practical issues that people must negotiate in social life. They are ripe for sociological analysis. Whereas the conceit of philosophers investigating epistemology is to find a universal grounding for the justification of knowledge, sociologists can bring their empirical sensibility to delineate how these epistemological issues get sorted in practice.

At root, this book serves as an exercise in the sociology of epistemologies (Abend 2006, 3), one that takes epistemologies as an object of analysis themselves and develops a conceptual toolkit that can assist in further research on the practices by which actors adjudicate true knowledge from false beliefs. It addresses a small, but burgeoning body of research that brings an empirically grounded sensibility to the study of epistemology. Research in historical epistemology challenges the misguided idea that there is a unitary timeless, universal epistemology against which all ideas must be measured by revealing how epistemological standards change over time (see Biagioli 1994; Daston 1992; Daston and Galison 2010; Davidson 2001; Dear 1992; Ginzburg 1980; Jonsen and Toulmin 1988; Poovey 1998; Schweber 2006; Shapin and Schaffer 1985). The fixed, universal standards, for which philosophers pine, are belied by the historical diversity in understandings as to what constitutes legitimate knowledge. Temporalizing the basic attributes of knowledge, historical epistemology disabuses us of the misconception that epistemological standards are timeless and that there exist epistemological rules that hold across all eras and contexts and justify all knowledge. Instead, it presents an invitation for analyses that situate epistemological claims within their historical period. This book shares a great affinity to this research. But it adds a focus on epistemological conflict *within* the same time period. Whereas historical epistemology focuses on comparison between eras, my work is attuned to struggles within an era of epistemological flux. Rather than focus on broad swaths of changes in knowledge, it attends to the strategies actors deploy when engaged in epistemological disputes during the same period, recognizing the importance of epistemological debates synchronically as well as diachronically.

Within sociology, the treatment of epistemology as an object of analysis is more scattered. While one can read the entire corpus of the sociology of science in the past three decades—including laboratory studies (e.g., Knorr-Cetina 1999; Latour and Woolgar 1986) and the sociology of knowledge downstream (e.g., Gieryn 1999)—as an epistemological challenge itself to the positivistic accounts of science (Longino 2002), much of the discussion around epistemology is circumscribed to the subfield's more reflexive moments, in metatheoretical arguments (Fuchs 1992, xvii). In other words, sociologists of science have spent so much time defending themselves against the indictment of relativism that they have not really turned to what should be their main concern: "how do people come to say they know things?" (Kurzman 1994).

Nevertheless, sociologists in a number of different subfields (e.g., economic sociology, political sociology, sociology of science, etc.) have taken up the challenge of approaching epistemology from an empirical standpoint. Although scattered and disparate, this research suggests rich areas to be explored further. Basic mental actions—classification, perception, the reckoning of time—have social foundations, shaped by the "epistemological styles" (Mallard, Lamont, and Guetzkow 2009) or "sociomental lenses" of the communities to which we belong (Zerubavel 1999). Indeed, conceptions of truth—and in turn, particular ideas—are developed within social networks, "not in isolated brains or disembodied minds" (Collins 2000, 877). Different national contexts produce divergent understandings of what constitutes legitimate knowledge, which is reflected in the diverse organizations for knowledge production (Fourcade-Gourinchas 2001) and the rhetorical forms in which knowledge claims are presented (Abend 2006). Laboratories themselves, though generally committed to the dictates of scientific norms, have diverse internal "epistemic cultures," which determine the nature of the knowledge they produce (Knorr-Cetina 1999). Organizational factors (e.g., task uncertainty and mutual dependence) determine the degree to which organizations allow for epistemic reflexivity (Fuchs 1992). And epistemological blind spots can produce accidents (Downer 2011) and undermine the effectiveness of political organizations (Glaeser 2011). While this body of research has yet to cohere into an organized research program, the shared sensitivity to the empirical analysis of epistemology presents many exciting opportunities for exploration.

This book builds on this body of research by introducing and developing the concept of the epistemic contest. By acknowledging the uniqueness of such knowledge disputes, it extends our understanding of the politics of knowledge while simultaneously opening a space for the analysis of epistemological change through struggle. The concept is offered as a complement to the rich tradition of sociological research on knowledge struggle. It is not intended to supplant and absorb this past research. In the end, the sociology of epistemologies that I call for in this book is situated within the tradition of the sociology of knowledge, not the other way around. Credibility contests still abound, as do other disputes that do not involve epistemic struggle. What I offer is a model for the examination of epistemic contests should the conditions of such be operative in a given case. But I want to caution those against reading every knowledge dispute as an epistemic contest. To be defined as an epistemic contest, disputes must revolve around fundamental epistemological issues. Whether these conditions hold is an empirical issue. If they do, then the insights of this book should be of great help in providing a framework and sensitizing concepts. If these conditions do not hold, then this book will offer some insights into other forms of knowledge disputes, but such use must, by necessity, be more careful and nuanced.

Ultimately, the usefulness of concepts and theories is that they allow the researcher to remain empirically responsible to the case under examination (Reed 2011). In introducing the concept of the epistemic concept, I hope to draw attention to the diversity in types of knowledge disputes, a diversity that is often obscured by the fact that sociological models of knowledge are built mostly upon the analyses of *scientific* debates. But knowledge disputes are not always about the boundaries of science; they can encompass more fundamental issues. We must appreciate the unique dimensions of epistemic contests and examine the strategies by which actors try to capture epistemic authority in knowledge disputes bereft of clear standards.

Though exceptional, epistemic contests address the most basic practical and social issue related to knowledge: how do we deem certain claims as legitimate and true, and others as illegitimate and false? Each chapter of the book elaborates on a particular facet of epistemic contests, elucidating the ways in which actors negotiate epistemological issues in practice. Epistemic contests are embedded in organizational arenas in which actors, advocating different epistemological systems, attempt to impose their vision of knowledge, gain epistemic authority, and achieve epistemic closure. The embeddedness of epistemic contests points to the importance of situated rhetoric and the mechanism of resonance in accounting for the success or failure of particular epistemological positions within a given setting. Because epistemic contests operate on a fundamental level in which standards are vague and ill-defined, they are more open than other types of knowledge disputes and, consequently, are waged with a great diversity of strategies. Indeed, perhaps the key difference between epistemic contests and other types of knowledge disputes is the extent to which actors deploy organizational strategies to gain an upper hand in what is, in its essence, a cultural dispute. The sociology of science downstream focuses on cultural practices like boundary work, but in those disputes, there is a well-defined cultural space—science—over which actors are fighting. When such a cultural space is absent, however, organizations become important arbiters in knowledge disputes, and actors thus seek to capture them to promote their epistemological visions. As this book shows, to win epistemic recognition, actors deploy diverse cultural and organizational strategies to frame epistemic authority in different ways , to construct and disseminate discoveries that validate their epistemological systems, and to win influential allies and resources.

Although these findings are born from a strident commitment to maintaining empirical fidelity to the particular case of nineteenth-century medical disputes, certain elements of my analysis can be abstracted to help understand other cases where similar conditions may hold. Here a brief word on generalizability is in order. Ultimately, the usefulness of any theory or concept is in its ability to interpret particular empirical cases. The recognition of specificity, contextuality, and contingency does not require forgoing general theoretical claims, although it does require some humility when making these claims. The strength of a single case study is that it allows for fine-grained, process-tracing that can accommodate complex causality and a more detailed examination of context (George and Bennett 2005). This is precisely what was needed to tell this story. By attending to the historical details of this particular case, I develop the concept of the epistemic contest by identifying important strategies, processes, and factors at play. But in adopting a single case study method, I trade depth for breadth. When it comes to other cases, there is a limitation to what I can say generally. *If* an epistemic contest is at play, then the facets discussed herein are operative, but whether or not these conditions hold, is, at the end of the day, an empirical question. What I offer future researchers is not a general theory of epistemic contests, but an elaboration of some of the various strategies, factors, and processes involved in epistemic contests identifiable through my case study. My goal has been to identify elements that may be more or less generalizable to other epistemic contests and that can provide a springboard for beginning to analyze how these issues play out in other debates and arenas. In other words, I hope that my analytical elaboration of the concept of epistemic contests provides some grist for the sociology of epistemologies mill.

Despite these limits, the idea of epistemic contests could prove quite fruitful for the examination of other cases. Subsequent scholars should take the concept up and manipulate it to fit the needs of their case. Here I offer only a few suggestions of the ways in which the concept might be useful.

First, as a general rule of thumb, intractable debates with incommensurable positions are a good place to find epistemic contests. Because medical issues involve the experiential knowledge of the patient, the practical knowledge of the clinician, and the scientific knowledge of the medical expert, medicine is rife with epistemic contests. For example, the incorporation of Complementary and Alternative Medicine (CAM) into mainstream scientific medicine involves some real epistemic tensions, especially around issues of testing and evaluating treatments. CAM advocates argue that biomedical testing procedures fail to capture the essence of CAM's efficacy. How can CAM treatments be adequately assessed if the whole of the person is not taken into account? Biomedical advocates retort that they cannot be expected to use unproven remedies. An uncertain stalemate ensues, bogged down in issues of incommensurability. Similar issues arise in cross-cultural exchanges of medical systems, particularly between Eastern and Western medical traditions. As medicine becomes globalized, these incommensurable systems interact, and often clash, in interesting ways. For example, in his analysis of the pluralities of modern Chinese medicine, Volker Scheid (2002) describes a hospital pharmacy in which one half is dedicated to modern medicines, the other to traditional herbal Chinese cures. One side is the model of modern science, the other the picture of folk wisdom. The two reside in the same space in a situation of constant tension and negotiation. Studying these separate-but-equal arrangements can shed light on epistemic contests that, unlike the case discussed in this book, do not result in one dominant epistemology or achieve epistemic closure. Andrew Lakoff (2005) has identified similar cross-cultural issues that have arisen in the importation of the biomedical model of mental illness to Argentina, a country with a long tradition of psychoanalytic psychiatry.

Second, the concept might help shed light on the divisions between biomedical research and clinical practice that are prevalent among health professionals (see Freidson 1970; Montgomery 2006). Misconstrued as a technical problem, the tension between researchers and clinicians has deep roots. Rather than the specific exigencies of clinical practice, the salient issue is epistemological in nature. Quite simply, the divergent roles that clinicians and researchers serve in the profession lead to different orientations toward knowledge and competing models of what constitutes useful knowledge. The epistemological tensions between researchers and clinicians reflect a classic distinction, noted by Aristotle, between *episteme* and *phronesis* (Jonsen and Toulmin 1988). *Episteme* is what we now understand as scientific reasoning in which the goal is to illuminate universal and general rules, to uncover timeless Truth. Clinicians, in contrast, approach knowledge differently, adopting a more practical posture toward knowing (Montgomery 2006). Practical wisdom, or what Aristotle calls *phronesis*, addresses particular cases and specific quandaries, employing, not maxims or rules, but a network of considerations to be tested by trial and error. It operates in the realm of the concrete, the temporal, and the presumptive. While *phronesis* and *episteme* are not inherently opposed, their relationship can be contentious, as the generalizing tendencies of *episteme* can threaten to devalue an appreciation of the idiosyncrasies of the clinical interaction. Understanding the epistemological roots in the researcher/clinician divide may prevent us from misreading issues like clinician resistance to evidence-based medicine (EBM) as merely technical problems and focus our attention on how to translate between epistemological orientations.

Finally, the concept could prove useful in examining those marginal diseases, or "contested diseases," like fibromyalgia, environmental illness, chronic fatigue syndrome, and sick-building syndrome (see Barker 2005; Dumit 2006; Kroll et al. 2000; Murphy 2006). Here the issue is a clash between the biomedical model and more experientially based knowledges. Unable to detect these diseases through conventional biomedical means, some medical professionals deny their existence. However, patient advocacy groups vie for recognition, and the resources (i.e., insurance coverage) that accompany recognition, by appealing to experiential knowledge. How these issues get sorted out in practice could be examined fruitfully by bringing these epistemological issues to the forefront of the analysis.

Epistemic contests are not confined to medicine; they proliferate in other areas of social life as well. The classic cases come from the perpetual disputes between religion and science. Indeed, the Catholic Church's censorship of Galileo and his heliocentric universe in the seventeenth century is in many ways the paradigmatic example of an epistemic contest. Galileo's insight was rooted in his telescope, the Church's in the Bible. Galileo drew his insight from his own eyes, the Church from its clerical tradition. The modern incarnation of this, the debates over evolution, repeats this age-old epistemic contest in many ways.[10] Other examples of contemporary epistemic contests are the debates over postmodernism in academia, local environmental disputes like controversies over fisheries that pit local experiential knowledge against environmental science (see Marlor 2010), conflicts between the collective memory of communities and the historical knowledge produced by historians (Whooley 2008), the various cross-cultural confusions that proliferate under globalization, and even talent evaluation in baseball front offices, where new, advanced statistical analyses clash with the traditional wisdom of the baseball scout (see Lewis 2004).

A final direction for future research into epistemic contests is to see whether the concept is useful when applied to cases at different levels of analysis via "analogical theorizing" (Vaughan 2004). While this book explores meso-level practices, there is no logical imperative to restrict epistemic contests to this level of analysis. Indeed, epistemic contests may be just as prevalent, or more so, at a micro-interactive level. Once again, medicine is a good place to find such micro-level epistemic contests. For example, there is research showing fundamental incongruity in understandings of mental illness that play out in the clinical interaction between psychiatrists and patients (Whooley 2010). Or one could examine the conflicts between the standards of scientific evidence and legal evidence in the courtroom through the lens of an epistemic contest (e.g., demanding an expert witness to claim 100 percent certainty when the standards of science cannot possibly allow her to do so). Such analyses might lack the ability to account for large-scale changes over time, but they would likely provide insight into the ways in which actors advocate for epistemological positions that are only visible at the interactive level.

Regardless of what form this research will assume, sociology stands to gain much insight by reorienting epistemology away from speculative thought experiments toward the everyday practices by which actors adjudicate knowledge claims. Epistemological issues are not just the province of ivory tower philosophers; they are practical problems that must be negotiated in social life.

Because of this, we can learn from the quacks, dreamers, and medical reformers of the nineteenth century.

APPENDIX

A COMMENT ON SOURCES

In researching this book, I examined a wide variety of historical documents, culling data from book manuscripts, professional journals, meeting minutes, newspapers, magazines, legislative documents, and even memoirs and diaries. I sampled documents from each of the relevant collective actors (i.e., orthodox physicians, homeopaths, Thomsonists, sanitarians) and organizations (i.e., state legislatures, professional societies, boards of health) involved in the epistemic contest. Using these source materials, I was able to index the changes in medical knowledge, reconstructing the history of cholera as an object of intellectual scrutiny so as to gain insight into the more general epistemic contest over medicine in the nineteenth century.

To provide a baseline for this history, I traced the debates over cholera as they took shape in two medical journals, the *Boston Medical and Surgical Journal* and the *Journal of the American Medical Association*. Not only do the pages of these two journals contain debates over cholera; they also include professional polemics as many authors used the journals as a forum to rebut claims of competing medical sects. Moreover, their continuity allowed me to follow the evolution of the debate for the entire time period under concern (roughly 1830 to 1915).

Having established this foundation, I conducted a more targeted investigation of additional archival materials, which included:

- *Public Documents of the Collective Actors Involved*. These documents—journal articles, pamphlets, and editorials—consciously address a public audience, and, thus, reveal actors' arguments, rhetoric, and frames regarding cholera, as well as the epistemic assumptions underlying these.
- *Meeting Records of the Collective Actors Involved*. Transactions of meeting proceedings by the relevant actors balance public documents by provid-

ing a window into actors' strategic deliberations, illuminating internal debates obscured in the more consciously public documents.
- *Institutional Records.* Institutional records (i.e., legislative documents, internal sanitary reports, etc.) provide insight into the efficacy of actors' arguments in particular organizational environments. Because epistemic contests occur within the institutions, institutional documents provide some sense as to the causes of success or failure of particular arguments. These documents include the institutional rationale for taking one side over another in the adoption of particular policies.

All told I analyzed and coded over one thousand documents in detail and read many, many more. I approached each document as a rhetorical object that reflected a particular position within the epistemic contest, situating it within the longer debate of which the document was a part. My analysis of the documents assumed a dual tack. First, when examining the debates over cholera, I attended to the manner in which actors made truth claims about the disease and understood the nature of medical knowledge, focusing on both the form and content of their arguments. I analyzed not only the specific claims about cholera but also the epistemological assumptions underlying these claims (i.e., the nature of the facts contained therein, the presentation of knowledge claims, the authorities drawn upon, etc.). Second, I drew on internal documents from relevant organizations to reconstruct the strategies that relevant collective actors employed in the epistemic contest. The epistemic contest did not only unfold on the pages of old medical journals; it encompassed a number of strategies, both cultural and organizational, by which actors sought to achieve epistemic recognition. In other words, these internal documents shed light on the professional strategies actors adopted. Furthermore, because institutions have their own epistemologies and cultural norms, I embedded my analyses of these strategies within their institutional context to account for their efficacy in specific institutions.

Most of these materials were located in seven archives: the Bobst Library at New York University, the Bradford Homeopathic Collection at the Taubman Health Science Library at the University of Michigan, the Butler Library at Columbia University, the New York Academy of Medicine, the New York Historical Society, the Parnassus Library at the University of California–San Francisco, and the Rockefeller Foundation Archive Center. Below is a list of key sources.

Orthodox Medicine

- *American Medical Times* (1860–1864)
- *Boston Medical and Surgical Journal* (1828–1928)
- *Journal of the American Medical Association* (1883–Present)
- *Medical and Surgical Reporter* (1856–1898)
- *New York Journal of Medicine and Collateral Sciences* (1843–1856)
- *Transactions of the American Medical Association* (1848–1882)
- *Transactions of the Medical Society of the State of New York* (1807–1925)
- *Transactions of the New York Academy of Medicine* (1851–1903)

Homeopathy

- *American Journal of Homeopathy* (1853–1923)
- *The Homeopathic Examiner* (1840–1847)
- *The Physician's and Surgeon's Investigator* (1880–1889)
- *Transactions of the American Institute of Homeopathy* (1844–1908)
- *Transactions of the Homeopathic Medical Society of the State of New York* (1863–1896)

Sanitarians

- *Annual Report of the Metropolitan Board of Health* (later titled *Annual Report of the Board of Health of the Health Department of the City of New York*) (1866–1912)
- *The Plumber and Sanitary Engineer* (1877–1880)
- *The Sanitarian*(1873–1904)
- *Selections from Public Health Reports and Papers Presented at the Meetings of the American Public Health Association* (1873–1907)

Thomsonism

- *Boston Thomsonian Manual* (1840–1845)
- *Botanico-Medical Recorder* (1837–1848)
- *The Thomsonian Botanic Watchman* (1834–1835)
- *The Thomsonian Messenger* (1841–1845)

Popular Press

- *Harper's Magazine* (1850–present)
- *The New York Times* (1851–present)

NOTES

INTRODUCTION

1. I refer to the dominant sect of nineteenth-century America as "regulars," "allopathic physicians," or "regular physicians." In adopting this nomenclature, I intentionally avoid referring to them as "orthodox physicians," a common practice among many historians. The use of "orthodoxy" tacitly privileges the dominant sect as "real" or "normal" medicine and conveys a degree of cohesion in thought that is historically problematic. This is not to suggest that my nomenclature is neutral. There are problems in using "regulars" and "allopathy" as well, namely the politically charged nature of the labels. Still, my choice of these two terms reflects an attempt at a more neutral nomenclature. The dominant sect referred to themselves as "regulars," while homeopaths and alternative medical sects (somewhat derisively) called them "allopaths." Using the two terms interchangeably I hope to, in effect, cancel out their ideological baggage. But there is no escaping the normatively laden nature of name-calling in nineteenth-century medicine. Recognizing this, I use "regulars" and "allopaths" heuristically to maintain the crucial incumbent/challenger distinction by identifying those in the dominant position in opposition to the challengers of alternative medical movements.

2. The miasmatic theory of disease, popular during the mid 1800s, held that diseases were caused by miasma (pollution), or noxious "bad air" in the atmosphere. Such air contained poisonous vapors and decaying material, that when inhaled, caused disease.

3. For example, Berlant (1975) examines how advocates of medical science came to monopolize medical work in the United States through the "organizational conduct" of AMA, which captured important organizations (i.e., hospitals) to achieve its goals (Berlant 1975, 48). Similarly, Freidson (1970) highlights the relation of medicine to the state to account for the great autonomy of the U.S. medical profession. Other studies focus on the importance of particular institutions for the consolidation of medical authority, such as hospitals (Rosen and Rosenberg 1983) and the educational system (Ben-David 1960; Ludmerer 1985; Markowitz and Rosner 1973).

4. This oversight could be avoided if Starr instead conceived of culture *as practice*—that is, culture as a sphere of practical activity shot through by willful action, power relations, struggle, contradiction, and change (Sewell 2005). By con-

ceiving culture as separate from, and indeed hovering above, human action, Starr fails to interrogate the ways in which cultural forms are produced and reproduced through the practical activities of social actors. Starr therefore tries to understand the transformation of medicine within the broader changes in culture and society, not the role that physicians played in *creating* these broader cultural changes.

5. Incidentally, Starr's temporality is a bit off, as the professionalization of medicine occurred *before* bacteriology's true successes. Furthermore, a valuation of scientific expertise under Progressivism, a key factor for Starr, did not dictate *which* scientific program was to be adopted. Most alternative medical sects saw their own programs as the epitome of science. The Progressive Era's general embrace of science cannot explain why the particular science of bacteriology was adopted.

6. This research resonates with the program of "social epistemology" as laid out by Steve Fuller (2002), which recognizes that knowledge is intrinsically social and that forms of knowledge contain an implied social order. However, Fuller situates his program of social epistemology squarely within philosophy; his intent is to achieve a normative assessment of the social organization of knowing. The analysis in this book resides more squarely within sociology; I seek to explain rather than prescribe. For this reason, I refer to what I'm doing as a "sociology of epistemologies" (Abend 2006, 3), not social epistemology. In adopting the plural "epistemologies" I intend to convey that epistemology is not *one* empirical object, the way philosophers conceive of it, but many different empirical objects.

7. Because of the ubiquity of the concept of paradigms in analysis of intellectual change, a word on the relationship between my analysis and Kuhn's is in order. Clearly, there are similarities between Kuhn's paradigm shifts and my exploration of epistemological change. They both challenge the progressive view of knowledge by showing how standards of good research change from era to era. The difference is one of emphasis. Kuhn focuses on scientific paradigms and models of good science that share more or less similar epistemological commitments (i.e., hypothesis testing, empiricism, etc.). I am interested in cases where these assumptions are contentious, arguing that the dynamics of epistemic contest vary in important respects from paradigm shifts.

8. In terms of knowledge debates, Abbott focuses primarily on the issue of abstraction and the various cognitive strategies professionals used to achieve "optimum abstraction" (Abbott 1988, 105). The case discussed in this book reveals a great diversity in types of strategies—cultural and organizational—deployed in epistemic contests.

9. Because of their ubiquity and authority, it is easy to forget that professions are a relatively new way to organize knowledge in society. It was not until the early twentieth century that they were institutionalized and became uncontroversial. Prior to this, the history of claims to professional standing were highly suspect. This history of professions is often obscured by sociologists, who tend to take professions for granted and, in the process, unintentionally naturalize them. The case under discussion in this book shows how, in the democratic fervor of the nineteenth-century United States, the legitimacy of professions was widely challenged.

10. Much of the explicit discussion of epistemology within SSK is relegated to reflexive moments, when SSK contrasts its research agenda with the dominant positivistic accounts of science or when defending itself against accusations of relativism (Kurzman 1994).

11. The great exceptions to this rule are feminist critiques of science that offer a more conscious exploration of epistemology. Donna Haraway's *Primate Visions* (2006) is an exemplar of this research, as she reveals how gendered thinking insinuates itself in the thinking and practices of primatologists.

12. While critical philosophical approaches (e.g., American pragmatism and naturalized epistemology) have tried to steer philosophy toward empirical analyses, it remains largely a speculative exercise, focused on elucidating logical arguments rather than studying *how* people negotiate these problems in real-life encounters.

13. The basic conceit of this book is that concept formation and elaboration are essential exercises in sociological research as good concepts "capture essences, identify dominant forces, determine our focus, and suggest future direction" (Light and Levine 1988,11).

14. Here I should acknowledge the limitations of a single case study. The strength of a single case study is that it allows for fine-grain process-tracing that can accommodate complex causality and a more detailed examination of context (George and Bennett 2005). My attention to historical detail allows for theory-building regarding epistemic contests by identifying important strategies, processes, and factors at play. But in adopting a single case study method, I trade depth for breadth. When extending the concept of epistemic contests to other cases, there is a limit to what I can say generally. I can say that *if* an epistemic contest is at play, then there are certain qualities it will assume, but whether or not these conditions hold, is, at the end of the day, an empirical question. What I offer future researchers is not a general theory of epistemic contests, but an elaboration of the various strategies, factors, and processes involved in epistemic contests.

15. This is not to suggest that Gieryn's work ignores epistemological issues altogether. To the contrary, he is concerned with "epistemic authority," which he defines as "the legitimate power to define, describe, and explain bounded domains of reality" (Gieryn 1999, 1), and some of his case studies—most notably, his analysis of John Tyndall's promotional work for science through London's Royal Institute—involved fundamental debates over the nature of knowledge. Nevertheless, these epistemological dimensions are not explicitly theorized by Gieryn. He is focused on investigating "science-in-culture," specifically how the adjudication of competing truths is accomplished through settlements of the boundaries of science. He wants to explore how and why science becomes so widely trusted. My analysis orbits around the more fundamental epistemological question of what constitutes legitimate knowledge, of which the answer "science" is only one of many possible answers.

16. While credibility contests originate in conditions in which science possesses an exalted status, historically the epistemological valuation of science fluctuates

over time. During the Jacksonian era, science was looked upon with great suspicion (Hofstadter 1963), and this suspicion fueled the epistemic contest over medicine.

17. The model of historical change animating my analysis is taken from Sewell (1992), who, while acknowledging the influence of structural factors on human action, retains a principal commitment to the role of the agency—the choices and understandings of historical actors—in bringing about transformations.

18. The counting of cholera epidemics is not as straightforward as one would think. For most of the epidemics discussed here, cholera stayed in the United States for a discrete period of time. This was not the case for the epidemic beginning in 1848. The intensity of the epidemic waned after 1849, but pockets of cholera remained throughout a six-year period, with an uptick in 1854, after which cholera disappeared for over a decade. Following the precedent set by Charles Rosenberg (1987b), I treat 1848, and its subsequent milder incarnations up to 1854, as a single epidemic.

19. In many ways, for the American South, yellow fever loomed larger than cholera as a medical issue (Crosby 2007; Ellis 1990; Humphreys 1999).

20. While I draw extensively on Rosenberg's research, my goals are different. Whereas Rosenberg describes how cholera became a secular *social* problem, I am interested in how it eventually became a *medical* problem to be understood through bacteriological science. This is reflected in the scope of my project. I examine all five cholera epidemics through 1892, while Rosenberg's analysis terminates in 1866.

21. While the best-fit cases were used as exemplars by proponents of the bacteriological theory, the assumption that one is compelled to adopt a single model of disease for all diseases is historically problematic. Indeed, long before the bacteriological revolution, the case of smallpox seemed to point toward the germ theory. Yet it was regarded by many to be a unique case. Similarly, John Snow's waterborne germ theory of cholera was presented not to offer a universal challenge to the miasma theory of disease, but rather to reveal cholera as an anomalous case that did not fit this model (Koch 2005). Part of the accomplishment of proponents of the germ theory of disease was convincing others that a single, universal paradigm explained all diseases. Reasoning by analogy undoubtedly was part of the rhetoric of proponents of the germ theory of disease, but there is no justification in arguing that a single case of success could have sealed the fate of alternative models.

22. It is a bit misleading to refer to the germ theory as a singular entity during the nineteenth century. As Worboys (2000) shows, there were many variations of the "germ theory," each with its own nuances. Nevertheless, for the purposes of this book, I refer to the germ theory to denote the belief in some sort of microbial etiology of disease. The different shadings of the germ theories are less important for my analysis.

CHAPTER ONE

1. Cholera had long circulated throughout India, but it was basically unknown to Europe prior to the 1800s.

2. This does not mean arenas deterministically predict outcomes. Social structures do not exist independently of human action. Per Giddens's (1984) structuration theory, James Jasper notes, "Behind every 'structure' is usually another player hard at work" (Jasper 2006, 167).

3. The use of the term "specificity" in nineteenth-century medical discourse is confusing and often contradictory, as historian John Harley Warner notes (1997, 58–63). During the early nineteenth century, regulars understood specificity as an individualized treatment for specific patients, and knowledge derived from observation of particular cases. This was in opposition to universalistic approaches to medicine. At the same time, they derided alternative medical movements and patent medicine sellers for championing specific remedies, as understood as specific to a particular disease, "as a manifest stigma of charlantry" (Warner 1997, 60). This confusing use of the term often results in a mischaracterization of nineteenth-century medical debates. For the purpose of this chapter, I employ the term in the former sense to signal a particularistic and local foundation for medical knowledge.

4. Lewins was onto something. Cholera essentially kills by extreme dehydration. Contemporary treatments involve oral rehydration therapy (ORT) or intravenous hydration, which both involve the prompt replacement of water and electrolytes. Unfortunately, amidst the myriad treatments for cholera during the nineteenth century, all of which promised wondrous results, the potentially effective treatment of saline injections failed to stand out.

5. My use of the feminine pronoun here is not done out of conformity to current intellectual standards. Given women's traditional role as nurturers in the household, Thomsonians targeted females as converts. Progressive in its gender politics, Thomsonism placed great emphasis in recruiting female members.

6. Hahnemann is also credited with inventing the label of "allopathic" medicine, which later became a term of derision that alternative medical sects hurled at regulars.

7. Contemporary scientific norms, such as those underlying the double-blind experimental design, may bias the modern reader against viewing self-experimentation as worthy of the status of science. However, there is a long, rich history of self-experimentation in medicine (Altman 1998).

8. New Jersey repealed its licensing laws much later in 1862.

9. Established in 1816, the Second Bank of the United States was chartered to enable the federal government to better manage its finances, particularly in the wake of spiraling inflation after the War of 1812. In the early 1830s, Andrew Jackson condemned the Bank as a bastion of fraud and vowed to "kill it." Jackson viewed the Bank as an abrogation of individual liberties and a gross overreach by the Federal Government that granted undue favor to bankers. The Bank loomed large as a controversy in American politics until 1836, when its charter expired.

10. The concept of elective affinity is used to illustrate the nondeterminative resonance between the practices of the state legislature and the democratized medical epistemologies of Thomsonians and homeopaths. This relationship is not determinative in that it took work and effort—mainly through their rhetori-

cal strategies—on the part of the alternative medical sects to draw these crucial connections. In other words, homologies do not automatically match up; links are *achieved* through the agency of actors.

11. For an extended discussion see Haller 2000, 131–138.

CHAPTER TWO

1. This particular story seemed to have great resonance, as it was repeated in so many different local contexts that it achieved almost mythical status.

2. Fortunately, some researchers have begun to address this oversight, bringing organizations back into the study of knowledge struggles. Recent research establishes the integral role that organizations can play in boundary work (Guston 1999; Kleinman and Kinchy 2003; Moore 1996). Organizations serve as important bridges between scientists and political actors, providing neutral spaces for the formation of networks across boundaries, all while maintaining the integrity of the separation between science and politics (Guston 1999; Moore 1996). This chapter identifies another, more basic role for organizations by identifying the ways in which organizations can be marshaled to promote particular epistemological agendas.

3. My use of the term "radical empiricism" is not meant to imply a relationship between mid-1800 regular medical epistemology and William James's (1976 [1912]) turn-of-the-century pragmatist philosophy, which also deploys the term. It is only intended to underscore the extreme degree to which regular reformers embraced empiricism and rejected all theory as illegitimate.

4. Given the intensity of the symptoms, this was not an impressive feat.

5. The AMA's success in achieving unity was limited by the geographical dispersion of allopathic physicians. Indeed, the society took on a regional character with most of the membership in its first few decades coming from five states: Massachusetts, Pennsylvania, New York, New Jersey, and Connecticut (Burrow 1963, 17).

6. Homeopaths were continuously frustrated by regulars' framing of their system as rationalism. "Their [allopaths'] mode of investigating Homeopathia is wrong. They assume that it is a mere theory, and because it is not, they of course cannot but misrepresent it. Now, if they would admit what is really true, that it is made up of facts arranged systematically for practical purposes, then the inquiry would be, are its proposed facts real facts, or otherwise? If Homeopathia is false, as allopaths declare, let them, in the legitimate way, answer that question. The answer, however, can be obtained only in one way, which is, to repeat Hahnemann's experiments" (Joslin 1852, 19–20).

7. The federal government was not too preoccupied to leave army physicians completely to their own devices. Union Surgeon General William Alexander Hammond, much to the chagrin of allopathic physicians, banned the use of the common allopathic treatments, calomel and tartar emetic. While homeopaths may claim some credit for this victory, it was the rotting jaws of soldiers suffering from mercury poisoning that led Hammond to act. Nevertheless, his decision was

viewed by allopaths as an attack upon their autonomy and as an explicit critique of their system.

CHAPTER THREE

1. Lumping these disparate actors under a single heading is a bit misleading as they shared little more than a commitment to urban sanitary reform. I use the label as a convenient shorthand, with the acknowledgment that the "sanitarians" were by no means a monolithic group with the cohesive identity of, say, the homeopaths.

2. In his classic work *On Airs, Waters and Places*, Hippocrates called for the close observation of local characteristics, based on the premise that disease is a product of specific locales.

3. Snow's map is often depicted as the crucial evidence in establishing the connection between the cholera contagion and contaminated water (Johnson 2006). Actually, the Snow map seems to have had minimal impact in the United States as it was rarely mentioned in the myriad discussions of cholera until after 1900 (after the germ theory became dominant). Snow is far better known now than he was in his lifetime (Hamlin 2009, 180). The idea that his map represented a singular discovery in the United States is a by-product of reading history backward.

4. Because dot maps fixed disease clusters in space, these maps contained little insight into temporality. Some of the more contagionist persuasion added dates to their dot maps to introduce a temporal dimension.

CHAPTER FOUR

1. Historians today acknowledge that an Italian doctor, Filippo Pacini, identified the bacillus forty years prior to Koch, so the attribution of discovery to Koch is not based on getting there first. According to historian William Coleman (1987, 330), this raises the issue of what it was that Koch accomplished: "What then, is in a 'discovery'? Obviously, the act of seeing the cholera vibrio was no novelty in 1884. And the association of this microbe with Asiatic cholera had also been made long before, and perhaps repeatedly."

2. Throughout this chapter, I refer to Koch's identification of the comma bacillus as a "finding," "idea," or "research." This reflects my claim that research becomes a discovery only *after* a long process of justifying it as such.

3. Latour has a tendency to depict these interests as fixed and preexisting; he never addresses their formation. For example, the hygienists, so crucial to Pasteur's success, are presented as an undifferentiated mass with obvious and enduring interests. Perhaps they were, but as this following case shows, collective actors are not necessarily homogenous and consistent in their interests. Given the heterogeneity of actors and motives, sense-making activities are essential to the success of a network.

4. Recent historical scholarship suggests Koch himself did not codify these postulates. Rather it was his colleague, Friedrich Löffler, who formally defined and popularized the postulates (Gradmann 2008). Nevertheless, Koch's own re-

search subscribed to them, albeit in a less systematic fashion than that in which they were subsequently laid out.

5. Koch later abandoned the healthy organism stipulation of the first postulate when he discovered asymptomatic carriers of cholera.

6. The logic behind vaccination was not new, as in the United States it dated back to the Puritan's use of inoculation for smallpox in the 1700s.

7. Given cholera's waterborne nature, it was probably not the best idea to dump it into a river.

CHAPTER FIVE

1. In the past, cholera had been associated with certain immigrants groups, most notably the Irish in 1832 (Kraut 1995).

2. A similar dynamic played out during an outbreak of the plague in San Francisco in first decade of the twentieth century, when anti-immigrant sentiment toward the Chinese shaped the city's response to the epidemic (Chase 2004). Nativism also arose during the 1892 typhus epidemic in New York with Russian Jews once again the scapegoats (Markel 1997).

3. A curious Rockefeller would later send specimens of her herbal remedies to the scientific laboratories he funded to see if they did indeed possess medicinal properties (Chernow 1998, 6).

4. Rockefeller was already sixty-two when the RIMR was founded.

5. In contrast to the minimal standards of proprietary schools, Hopkins expected much more of its students, prior to their arrival, during their stay, and after their graduation. Its other innovations included the requirement of a bachelor of arts for admission, acceptance of female students, a four-year graded curriculum, and a grading system for interns and residents.

6. While the Flexner Report looms large in the historical imagination as a cause of medical educational reforms, recent histories offer a more measured analysis, viewing it as part of the broader movement of educational and scientific reforms (e.g., Banta 1971; Ludmerer 1985).

7. These costs were hard to bear even for hospitals, as the expense of laboratories strained already tight budgets (Rosenberg 1987a). Still, by pooling resources and consolidating practices, hospitals adjusted.

8. Homeopathy was defeated but not dead. It experienced a rebirth in the 1960s and 1970s, as scientific medicine became criticized for its dehumanization of the patient. However, the new incarnation differed drastically from its nineteenth-century counterpart, as it embraced its more mystical side in an attempt to offer a holistic alternative to mainstream medicine.

CONCLUSION

1. De Kruif would later publish his own history of the early decades of the germ theory, entitled *Microbe Hunters* (1926), a romantic, best-selling hagiography that

would form the foundation of the conventional truth-wins-out narrative of the emergence of laboratory medicine.

2. Not all literature celebrated these developments in medical science. The counterpoint to Arrowsmith, the villain to this laboratory hero, was Georg Letham, the main character in Ernst Weiss's (2009) novel *Georg Letham: Physician and Murderer*. Weiss, a doctor himself writing in the epicenter of bacteriology—Austria and Germany—was suspicious of the new laboratory sciences and the model of medicine it promoted. His novel recounts the dark story of Dr. Georg Letham, who murders his patients, via bacteriological infection, in order to advance his scientific research on yellow fever. The novel presages (albeit in an extreme way) later critiques that laboratory sciences dehumanized medicine. And written in Austria in 1931, it is an eerie premonition of the horrific experiments that would be conducted by Nazi scientists a decade later.

3. Gottlieb was modeled after Frederick D. Novy, a bacteriologist, and Jacques Loeb, an American physiologist and experimental biologist at the Rockefeller Institute. Both were heroes and mentors to de Kruif (Rosenberg 1963).

4. Fourcade (2009) describes a similar process in twentieth-century economics, demonstrating how the particular political, cultural, and institutional contexts in different countries gave rise to distinct professional and disciplinary configurations.

5. In his history of early attempts on the part of reformers to achieve government-led health insurance, historian Ronald Numbers (1978) argues that the AMA initially entertained the idea and were "almost persuaded" to accept such reforms. Eventually, according to Numbers, the red-scare politics of World War I coupled with resistance by local medical societies led to the defeat of such efforts. Numbers's provocative title, *Almost Persuaded*, obscures what is a much more ambiguous history. As even his own evidence shows, the AMA was always thoroughly ambivalent toward such plans. The sum total of its engagement with the reform efforts—aside from a few individuals—was to establish a committee to investigate the issue, and it did this only when such reforms appeared to be inevitable (especially in the wake of the establishment of government health care insurance in Germany and Britain). It is telling that the only official declaration the AMA made during this period came in 1921, when it publicly rejected such plans. Rather than "almost persuaded," the AMA's initial forays into investigating government-led health insurance represented an attempt to hedge its bets. Once the perceived inevitability of such reforms gave way, the AMA showed its true colors by actively working to defeat them—a position consistent with the strategic orientation it embraced in the aftermath of the epistemic contest.

6. Jamie Ferran's vaccine (discussed in chapter 5) was beset with problems and the international medical community was reluctant to embrace his conclusions (Bornside 1982). In 1892, Waldemar Haffkine developed an anticholera vaccine that was looked upon quite favorably by the international medical community. However, after a number of missteps and controversies, enthusiasm waned and the vaccine never caught on (Löwy 1992).

7. Vaccination actually preceded laboratory science as the smallpox vaccine was developed by Edward Jenner in 1796. Even earlier, during the colonial period in the United States, Cotton Mather suggested that inhabitants of the Massachusetts Bay Colony could get inoculated against small pox (Silverman 1984).

8. Although Joseph Lister published his famous paper "On the Antiseptic Principle of the Practice of Surgery" in 1867, the mechanisms behind these techniques were later made explicit by Louis Pasteur. As the germ theory gained widespread acceptance, such techniques were refined, leading to the rise of sterile surgery.

9. The tension between democracy and science identified here diverges from Robert Merton's (1973) understanding of the complementary and reinforcing relationship between democracy and science. Merton's claims emerge from a comparison between Western democracies and the restrictive practices of authoritarian regimes. This comparative lens leads Merton to stress the ways in which democratic social orders promote the scientific norms of value neutrality (or disinterestedness) and unrestricted rational and open discussions (or universalism). The key assumption underlying his claim is that science is open to all. While this may be formally true, I am stressing the barriers that are erected to prevent widespread participation, which retard democratic participation not just in the production of scientific knowledge but also in policy issues in which science is involved. As always Merton's actual analysis is more nuanced than the caricature of it, and he was cautious in not overstating the positive relationship between science and democratic social orders (Sica 2010). Still, we must recognize the ways in which science clashes with democratic values.

10. However, one needs to be careful, as the recent adoption of Intelligent Design (ID) by U.S. conservatives may suggest an epistemic convergence; to be acceptable creationism has been recast in terms of science. The degree to which ID represents a genuine effort at science or a cynical Trojan horse is an empirical question that will determine whether thinking of this issue as an epistemic contest is useful.

REFERENCE LIST

PRIMARY SOURCES

"An Act." March 4, 1837. *Southern Banner*, 1.

"Best Preparation for the Cholera." 1849. *New York Evangelist* 20, no. 24: 94.

"The Cholera at Pittsburgh." 1854. *German Reformed Messenger* 20, no. 4: 4166.

"Cholera Incident." 1849. *Liberator* 19, no. 31: 124.

"Cholera Voice." 1832. *Boston Medical and Surgical Journal* 6, no. 9: 148.

"A City in Mourning." 1849. *Christian Advocate Journal* 24, no. 30: 119.

"Death of Doctor Thomas Spencer." June 6, 1857. *Daily Courier*, 6.

"Malt Liquors and the Cholera." 1849. *National Era* 3, no. 32: 125.

"Rockefeller's Institution for Medical Research." 1901. *Medical Record*, 907. Rockefeller Institute Archives, Record Group 2, Box 52, Folder 539.

"A Sad Story—Effect of the Will." 1849. *Liberator* 19, no. 31: 124.

"Street Commissioners to the Cholera." November 9, 1865. *Nation*, 583–584.

Agnew, C. R. 1874. "Presidential Opening Remarks." *Transactions of the Medical Society of the State of New York, 1873–1874*, 4–47. Albany, NY: Van Benthuysen and Sons.

Allen, J. A. September 5, 1832. "Epidemics: Remarks on the Etiology and Character of Epidemics." *Boston Medical and Surgical Journal* 7, no. 4: 53–55.

AMA (American Medical Association). 1850. "Report of the Committee on Practical Medicine and Epidemics." *Transactions of the American Medical Association*, 107–130. Philadelphia: T. K. and P. G. Collins.

———. 1851. *Code of Medical Ethics of the American Medical Association*. Chicago: American Medical Association Press.

———. 1892a. Protective Vaccination against Cholera." *Journal of the American Medical Association* 19, no. 18: 529–530.

———. 1892b. "Laboratory Work in Medical Schools." *Journal of the American Medical Association* 19, no. 4: 110–111.

———. 1892c. "Quarantine." *Journal of the American Medical Association* 19, no. 15: 442–443.

———. 1892d. "Results of Researches in Bacteriology." *Journal of the American Medical Association* 19, no. 26: 757.

American Medical Times. 1860. "Our Sanitary Defenses."*American Medical Times* 1: 46–47.

———. 1862. "Relation of the Sanitary Condition of New York to the Country." *American Medical Times* 4: 98–99.

———. 1863. "The Week." *American Medical Times* 6: 59.

Atkins, Dudley. 1832. *Reports of Hospital Physicians and Other Documents in Relation to the Epidemic Cholera of 1832*. New York: G. and C. and H. Carvill.

Barker, Fordyce. 1860. "Annual Address." *Transactions of the Medical Society of the State of New York, 1860*, 5–10. Albany, NY: Charles van Benthuysen.

Bartlett, Elisha. 1844. *An Essay on the Philosophy of Medical Science*. Philadelphia: Lea and Blanchard.

Bartley, Horatio. 1832. *Illustrations of Cholera Asphyxia in Its Different Stages, Selected from Cases Treated at the Cholera Hospital, Rivington Street*. New York: S. H. Jackson.

Bates, Joseph. 1849. "The Resources of the Medical Profession." *Transactions of the Medical Society of the State of New York, 1847, 1848, 1849*, 17–34. Albany, NY: J. Munsell.

Bigelow, Henry Jacob. 1871. *Medical Education in America: Being the Annual Address Read before the Massachusetts Medical Society, June 7, 1871*. Cambridge, MA: Welch, Bigelow.

Billings, Frank Seaver. 1885. "Cultivations of the Cholera Bacillus and of Other Allied Microorganisms." *Boston Medical and Surigical Journal* 112, no. 20: 476.

Billings, John S. 1879. "The Study of Sanitary Science." *Plumber and Sanitary Engineer* 2, no. 5: 125.

Bissell, D. P. 1864. "Annual Address." *Transactions of the Medical Society of the State of New York*, 3–23. Albany, NY: Van Benthuysen and Sons.

Blake, E. 1894 "Cholera: Its Prevention and Treatment." *Transactions of the American Institute of Homeopathy* 4, no. 7: 880–889.

Blatchford, Thomas W. 1852. "Homeopathy Illustrated." *Transactions of the Medical Society of the State of New York, 1850, 1851, 1852*, 69–141. Albany, NY: Weed, Parsons, Public Printers.

Bossey, P. 1832. "Comparative Treatment of Cholera." *Boston Medical and Surgical Journal* 7, no. 16: 245–247.

Boston Medical and Surgical Journal (BMSJ). August 16, 1831a. "The Nature and Cure of the Indian Cholera." *Boston Medical and Surgical Journal* 5, no. 1: 5–17.

———. October 25, 1831b. "The Cholera and Its Treatment." *Boston Medical and Surgical Journal* 5, no. 12: 170–174.

———. May 2, 1832a. "The Treatment of Epidemic Cholera." *Boston Medical and Surgical Journal* 6, no. 12: 189–191.

———. May 30, 1832b. "The Cholera." *Boston Medical and Surgical Journal* 6, no. 16: 254–255.

———. July 4, 1832c. "Massachusetts Report on Cholera." *Boston Medical and Surgical Journal* 6, no. 21: 337–340.

———. July 11, 1832d. "Cholera at New York." *Boston Medical and Surgical Journal* 6, no. 22: 353–356.

———. November 28, 1832e. "The March of Cholera." *Boston Medical and Surgical Journal* 7, no. 16: 253–254.

———. June 5, 1833a. "After Thoughts on Malignant Cholera." *Boston Medical and Surgical Journal* 8, no. 17: 271–273.

———. June 26, 1833b. "Remarks on Cholera." *Boston Medical and Surgical Journal* 8, no. 20: 314–316.

———. August 12, 1835. "Remarks on Cholera." *Boston Medical and Surgical Journal* 13, no. 1: 13–14.

———. May 11, 1842. "Case of Death from Thomsonism." *Boston Medical and Surgical Journal* 26, no. 14: 216–218.

———. September, 12, 1849. "Cholera—Its Course and Ravages." *Boston Medical and Surgical Journal* 41, no. 6: 123–124.

———. January 15, 1885. "Demonstration of Koch's Bacilli Cholera." *Boston Medical and Surgical Journal* 111, no. 25: 108.

Boston Thomsonian and Lady's Companion (BTLC). 1840. "Regular Quackery." *Boston Thomsonian and Lady's Companion* 6, no. 22: 338–340.

———. 1841. "Simplicity." *Boston Thomsonian and Lady's Companion* 7, no. 8: 114.

Boston Thomsonian Manual. 1841. "Thomsonism." *Boston Thomsonian Manual* 7, no. 7: 98.

Bowers, B.F. 1871. "Anniversary Address." *Transactions of the Homeopathic Medical Society of the State of New York, 1871*, 101-126. Albany, NY: The Argus Company, Printers.

———. 1868. "Opposition to Homeopathy in New York." *Transactions of the Homeopathic Medical Society of the State of New York for the Year 1868*, 393–412. Albany, NY: The Argus Company, Printers.

Brinsmade, T. C. 1859. "Annual Address before the Medical Society and Members of the Legislature." *Transactions of the Medical Society of the State of New York, 1859*, 5–30. Albany, NY: Charles van Benthuysen.

Bronson, Henry. 1832. "Remarks on the Chlorides and Chlorine." *Boston Medical and Surgical Journal* 7, no. 6: 85–95.

Butler, S. W., R. J. Levis, and L. C. Butler. 1861. "Mr. Fergusson Holding the Professional Intercourse with Homeopaths." *Medical and Surgical Reporter* 62, no. 2: 496–497.

Carnegie, Andrew. 1889. *The Gospel of Wealth*. London: F. C. Hagen.

Chapin, Charles V. 1934a. "Justifiable Measures for the Prevention of the Spread of Infectious Diseases." In *The Papers of Charles V. Chapin*, 76–91. New York: The Commonwealth Fund.

———. 1934b. "Effective Lines of Health Work." In *The Papers of Charles V. Chapin*, 37–45. New York: The Commonwealth Fund.

———. 1934c. "Dirt, Disease, and the Health Officer." In *The Papers of Charles V. Chapin*, 20–27. New York The Commonwealth Fund.

Chapman, N. 1848. "Address." *Transactions of the American Medical Association*, vol. 1, 7–8. Philadelphia: T. K. and P. G. Collins.

Citizens' Association of New York. 1866. *Report of the Council of Hygiene and Public Health of the Citizens' Association of New York upon the Sanitary Condition of the City*. New York: Appleton.

Clark, A. 1853. "Annual Address Delivered before NY State Medical Society and Members of the Legislature." *Transactions of the Medical Society of the State of New York, 1853*, 271–295. Albany, NY: Charles van Benthuysen.

Clarke, Charles. 1846. "On Cholera, Its Nature and Treatment." *Boston Medical Surgical Journal* 35, no. 1: 9–11.

Colby, B. 1839. "Thomsonian Lecture." *Boston Thomsonian and Lady's Companion* 6, no. 1: 1–2.

Comstock, Joseph. 1832. "The Causes of Epidemics." *Boston Medical Surgical Journal* 7, no. 10: 149–159.

Cornell, Benjamin F. 1868. "Presidential Address." *Transactions of the Homeopathic Medical Society of the State of New York for the Year 1868*, 3–4. Albany, NY: The Argus Company, Printers.

Deloney, Edward. 1835. "Quackery." *Boston Medical and Surgical Journal* 7, no. 7: 111–112.

de Tocqueville, Alexis. 2000. *Democracy in America*. New York: Perennial Classics.

Dick, Robert. 1849. "A Few Observations on Cholera." *Boston Medical and Surgical Journal* 41, no. 3: 1–4.

Dickens, Charles. 2000. *American Notes for General Circulation 1842*. Edited by P. Ingham. London: Penguin Press.

Dickson, S. H. 1849. "On the Progress of Asiatic Cholera during the Years, 1844–46–47–48." *New York Journal of Medicine* 2, no. 2: 9–20.

Dixon, Archibald. 1885. "Progress in Medicine." *Journal of the American Medical Association* 5, no. 15: 416–417.

Eliot, Charles W. 1896. "The Medical Education of the Future." *Transactions of the Medical Society of the State of New York, 1896*, 87–104. Philadelphia: Dornan Printer.

Fitz, R. H. 1885a. "The Recent Investigations concerning the Etiology of Cholera." *Boston Medical and Surgical Journal* 112, no. 8: 169–172.

———. 1885b. "The Recent Investigations concerning the Etiology of Cholera." *Boston Medical and Surgical Journal* 112, no. 9: 196–199.

Flexner, Abraham. 1910. *The Flexner Report on Medical Education in the United States and Canada*. New York: Carnegie Foundation for the Advancement of Learning.

Flexner, Abraham, and James Thomas Flexner. 1941. *William Henry Welch and the Heroic Age of American Medicine*. New York: The Viking Press.

Flint, Austin. 1884. "Parasitic Doctrine of Epidemic Cholera." *Boston Medical and Surgical Journal* 111, no. 18: 420–423.

Gates, Frederick T. 1911a. "Letter to John Rockefeller, Sr. January 17, 1911." Frederick T. Gates Papers. Rockefeller Institute Archives Record Group FTG, Box 2, Folder 33.

———. 1911b. "Letter to Starr Murphy to Rockefeller, May 16, 1911." John D. Rockefeller, Senior Correspondence. Rockefeller Institute Archives Record Group 2, Box 29, Folder 228.

———. 1964. "Recollections of Frederick T. Gates on the Origin of the Institute." In *A History of the Rockefeller Institute*, edited by George W. Corner, 575–584. New York: Rockefeller Institute Press.

Grabill, J. B. 1857. *Petition of Physicians and Surgeons to the Legislature of this State in Favor of Introducing the Homeopathic System of Medical Treatment into the Insane Asylums, Hospitals, Prisons and Public Institutions of the State*. Union City, IN: Times New and Job Printing Office.

Griscom, John H. 1857. "Improvements of the Public Health, and the Establishment of a Sanitary Police in the City of New York." *Transactions of the Medical Society of the State of New York, 1855, 1856, 1857*, 107–123. Albany, NY: Charles van Benthuysen.

Halley, William. 1887. "The Plumbers' Plea for Representation on Boards of Health." *Sanitarian* 18, no. 208: 241–246.

Hamilton, Frank H. 1884. "The Asiatic Cholera at Suspension Bridge in 1854 and Its Lessons—What We Know about Cholera." *Boston Medical and Surgical Journal* 111, no. 21: 491–493.

Hand, S. D. 1874. "Proceedings of the 22nd Annual Meeting." *Transactions of the Homeopathic Medical Society of the State of New York for the Year 1874*, 1–28. Albany, NY: The Argus Company, Printers.

Harris, Elisha. 1869. "Report on Sanitary Police Applied to the Prevention and Control of Epidemic Cholera." *Bulletin of the New York Academy of Medicine*, vol. 3, 102–123. New York: William Wood.

Hildreth, E. A. 1868. "Report on the Epidemic Diseases of West Virginia." *Transactions of the American Medical Association* 14: 211–237.

Hiller, Frederick. 1867. *Medical Truth and Light for the Million*. Virginia, NV: Territorial Enterprise Book and Job Printing.

Holmes, Oliver Wendell. 1842. *Homeopathy and Its Kindred Delusions: Two Lectures Delivered before the Boston Society for the Diffusion of Useful Knowledge*. Boston: William D. Ticknor.

Homeopathic Medical Society of the State of New York. 1866a. "Annual Meeting of the Kings County Homeopathic Medical Society." *Transactions of the Homeopathic Medical Society of the State of New York for the Year 1866*, 241–251. Albany, NY: The Argus Company, Printers.

———. 1866b. "Fourth Annual Report." *Transactions of the Homeopathic Medical Society of the State of New York for the Year 1866*, 131–132. Albany, NY: The Argus Company, Printers.

———. 1866c. "Homeopathy and the Board of Health." *Transactions of the Homeopathic Medical Society of the State of New York for the Year 1866*, 320–323. Albany, NY: The Argus Company, Printers.

———. 1868. "Seventeenth Annual Meeting." *Transactions of the Homeopathic Medical Society of the State of New York for the Year 1868*, 3–28. Albany, NY: The Argus Company, Printers.

———. 1874. "Resolution." *Transactions of the Homeopathic Medical Society of the State of New York for the Year 1874*, 27–28. Albany, NY: The Argus Company, Printers.

———. 1910. "Resolution Regarding the Flexner Report." *Transactions of the Homeopathic Medical Society of the State of New York for the Year 1910*, 215. Rochester, NY: Rochester Democratic and Chronicle.

Hooker, Worthington. 1849. *Physician and Patient*. New York: Arno Press.

———. 1852. *Homeopathy: An Examination of Its Doctrines and Evidences*. New York: Charles Scribner.

Hott, Moore. March 7, 1832. "Remarks on Cholera." *Boston Medical and Surgical Journal* 6, no. 4: 59–64.

Hull, Lawrence. 1840. "Annual Address Delivered before the Medical Society of the State of New York, Feb. 6, 1839." *Transactions of the Medical Society of the State of New York, 1838, 1839, 1840*, 59–70. Albany, NY: J. Munsell.

Hun, Thomas. 1863. "Influence of the Progress of Medical Science over Medical Art." *Transactions of the Medical Society of the State of New York, 1863*, 3–36. Albany, NY: Charles van Benthuysen.

Hutchinson, J. C. 1867. "Annual Address." *Transactions of the Medical Society of the State of New York, 1867*, 53–69. Albany, NY: Charles van Benthuysen.

Jacobi, Abraham. 1885. "Inaugural Address." *Medical Recorder* 26: 169–174.

———. 1897. "Hygiene and Its Accessories." *Sanitarian* 38, no. 336: 385–412.

Jones, Daniel T. 1861. "Annual Address before Medical Society and State Legislature." *Transactions of the Medical Society of the State of New York, 1861*, 5–15. Albany, NY: Charles van Benthuysen.

Joslin, B. F. 1852. "Address." *Proceedings of the American Institute of Homeopathy*, vols. 4–9, 9–24. New York: Charles G. Dean.

Kellogg, Edwin Merrill. 1872. *Statistics of the Comparative Mortality of New York City, during the Years 1870 and 1871*. New York: Stearns and Beale.

King, Dan. 1849. "The Evils of Quackery and Its Remedies." *Boston Medical and Surgical Journal* 40, no. 19: 370–376.

Kinsman, D. N. 1886. "Etiology and Prophylaxis of Cholera Asiatica." *Sanitarian* 161, no. 99: 525–538.

Knight, J. 1846. "An Address to the Medical Profession, in Relation to the Objects of the National Medical Association by the Committee Appointed for that Purpose." *The Medical Examiner*, vol. 2, 748–751. Philadelphia, PA: Lindsay and Blakiston.

Lewins, Robert. 1832. "Injection of Saline into the Veins." *Boston Medical and Surgical Journal* 6, no. 24: 373–375.

Lippe, Arthur. 1865. *Who Is a Homeopathician?* Philadelphia: King and Baird Printers.

Lister, Joseph. 1867. "On the Antiseptic Principle in the Practice of Surgery." *British Medical Journal* 2, no. 351: 246–248.

Loomis, Alfred L. 1888a. "Address in Practice of Medicine, Materia Medica, and Physiology." *Transactions of the American Medical Association*, vol. 29, 119–137. Philadelphia: Collins Printer.

———. 1888b. "Anniversary Address." *Transactions of the Medical Society of the State of New York, 1888*, 52–71. Syracuse, NY: The Syracuse Journal Co.

Lynde, John S. 1848. "Who Shall Decide When Doctors Disagree?"*Maine Farmer* 16, no. 49: 2.

Macneven, William H. 1849. "Remarks on the Mode by Which Cholera Is Propagated." *New York Journal of Medicine* 2, no. 2: 186–209.

Maddux, D. P. 1892. *Significance of Bacteriological Discoveries to the Homeopathic Method of Treatment*. Reprinted in *Hahnemannian Monthly*.

Magendie, M. 1832. "Report from M. Magendie on His Trip to Sunderland, England." *Cholera Gazette* 1, no. 1: 6.

Markham, H. C. 1888, "State Regulation of the Practice of Medicine—Its Value and Importance." *Journal of the American Medical Association* 10, no. 1: 5–7.

McClelland, James Henderson. 1908. *Homeopathy, A System of Rational Therapeutics: Its Right to Survive*. Scranton: Homeopathic Medical Society of Pennsylvania.

Medical and Surgical Reporter. 1865. "Letter from Toger." *Medical and Surgical Reporter* 12, no. 1: 214–216.

———. 1866a. "The Metropolitan Board of Health and the Academy of Medicine; A Point of Ethics." *Medical and Surgical Reporter* 14, no. 24: 474 –477.

———. 1866b. "Homeopathy and Public Hygiene." *Medical and Surgical Reporter* 14, no. 5: 95–96.

———. 1867. "Cholera and Its Prevention." *Medical and Surgical Reporter* 17, no. 1: 16–17.

———. 1892. "Recent Investigations regarding the Aetiology and Toxicology of Asiatic Cholera." *Medical and Surgical Reporter* 66, no. 1: 459–462.

———. 1893. "Cholera This Season." *Medical and Surgical Reporter* 68, no. 11: 421–422.

Medical Society of the State of New York. 1867a. "Report." *Transactions of the Medical Society of the State of New York, 1867*, 65–69 Albany, NY: Van Benthuysen and Sons.

———. 1867b. "Sixtieth Annual Meeting." *Transactions of the Medical Society of the State of New York, 1867*, 4–43. Albany, NY: Van Benthuysen and Sons.

———. 1870. *Transactions of the Medical Society of the State of New York, 1870*. Albany, NY: Van Benthuysen and Sons.

Metcalf, __. 1869. "Discussion on Epidemic Cholera." *Bulletin of the New York Academy of Medicine* 3, no. 1: 1–25.

Metropolitan Board of Health. 1866. *Annual Report of the Metropolitan Board of Health*. Vol. 1. New York: C. S. Westcott Union Printing House.

———. 1867. *Annual Report of the Metropolitan Board of Health*. Vol. 2. New York: Union Printing House.

Miller, T. Clarke. 1887. "The Relation of the Physician to Sanitation." *Sanitarian* 19, no. 213: 107–144.

M'Naughton, James. 1852. "Annual Address Delivered before the Albany County Medical Society." *Transactions of the Medical Society of the State of New York, 1850, 1851, 1852*, 126–143. Albany, NY: Weed, Parsons Public Printers.

Munger, E. A. 1865. "Presidential Address." *Transactions of the Homeopathic Medical Society of the State of New York for the Year 1865*, 22–25. Albany, NY: The Argus Company, Printers.

Nation. 1865. "The Street Commissioners to the Cholera." *Nation* 1, no. 19: 583–584.

———. 1866. "New York and the Cholera." *Nation* 2, no. 28: 40–41.

Newman, James A. 1856. "Report on the Sanitary Police of Cities." *Transactions of the American Medical Association*, vol. 9, 429–482. Philadelphia: T. K. and P. G. Collins.

New York Assembly Select Committee on Petitions. 1842. "Report of the Select Committee on Sundry Petitions for a Law to Enable Thomsonian Physicians to Collect Pay for Their Services, &c." *Documents of the Assembly of the State of New York* 2, no. 142: 1–4.

———. 1843. "Report of the Select Committee on Petitions, for the Repeal of Laws Restricting Medical Practice." *Documents of the Assembly of the State of New York* 3, no. 62: 1–8.

———. 1844. "Report of the Select Committee on Petitions, Praying for the Repeal of Laws Restricting Medical Practice." *Documents of the Assembly of the State of New York* 4, no. 60: 1–4.

New York Journal of Medicine. 1844a. "Laws of New York Relative to the Practice of Physic and Surgery." *New York Journal of Medicine* 3, no. 1: 281–287.

———. 1844b. "Nature and History of Vital Statistics." *New York Journal of Medicine* 1, no. 3: 320–334.

———. 1849a. "Analysis of the Treatment of Cholera by Several Writers." *New York Journal of Medicine* 2, no. 3: 73–78.

———. 1849b. "The Progress of Cholera in America." *New York Journal of Medicine* 2, no. 2: 97–99.

New York Senate Select Committee on Petitions. 1844. "Report of the Select Committee on Petitions for the Repeal of Laws Restricting Medical Practice." *Documents of the Senate of the State of New York* 1, no. 31: 1–4.

New York Times. June 25, 1856. "Killing Off Our Children—By Authority." *New York Times*, 3.

———. April 9, 1866. "The Approach of Cholera." *New York Times*, 4.

———. July 1, 1866. "The Academy of Medicine and the Cholera." *New York Times*, 4.

———. July 19, 1866. "The Public Health." *New York Times*, 2.

———. February 20, 1867. "Medical Rivalries and Claims." *New York Times*, 4.

———. March 31, 1867. "Cholera in New-York in 1866." *New York Times*, 3.

———. October 28, 1883. "Progress of the Germ Theory." *New York Times*, 8.

———. September 9, 1892. "The Record Not So Good." *New York Times*, 1.

NYAM (New York Academy of Medicine). 1862. "Resolution on the Inclusion of Homeopaths in the Union Army,. *Transactions of the Medical Society of the State of New York, 1862*, 435. Albany, NY: Charles van Benthuysen.

———. April, 28, 1866. "Minutes of the New York Academy of Medicine." *New York Academy of Medicine Archives*.

———. 1871. "Discussion on Cholera." *Transactions of the New York Academy of Medicine*, vol. 22, 1–47. New York: Bailliere Brothers.

Orme, Francis H. 1868. *Homeopathy—What Is It?* Detroit, MI: E. A. Lodge.

Osler, William. 1895. *The Principles and Practice of Medicine*. 2nd ed. New York: D. Appleton.

Page, Fred. 1849. "Remarks on Epidemic Cholera." *Boston Medical and Surgical Journal* 40, no. 22: 1–11.

Paine, Horace. 1866. "Epidemic Cholera." *Proceedings of the American Institute of Homeopathy*, vol. 19, 126–144. Boston: Rand and Avery.

Pintard, John 1832. *Letters from John Pintard to His Daughter Eliza Noel Pintard Davidson*. Vol. 4, 1832–1833. New York: J. J. Little and Ives.

Plumber and Sanitary Engineer. 1879. "Sanitary Science." *Plumber and Sanitary Engineer* 2, no. 4: 94.

Prudden, T. Mitchell. 1889. *The Story of Bacteria*. New York: G. P. Putnam's Sons.

Quackenbush, J. V. 1869. "Introductory Remarks." *Transactions of the Medical Society of the State of New York*, 4–5. Albany, NY: Van Benthuysen and Sons.

Reese, David Meredith. 1833. *A Plain and Practical Treatise on the Epidemic Cholera, as It Prevailed in the City of New York, in the summer of 1832*. New York: Conner and Cooke.

Roe, John O. 1899. "The Relation of Medicine to Civilization." *Transactions of the Medical Society of the State of New York*, vol. 51, 44–61. Philadelphia: Dornan Printer.

Rose, John. October 1838. "Carroll County Thomsonian Emancipation Candidates." *Lobelia Advocate* 1: 10.

Sanitarian. 1873. "Cholera." *Sanitarian* 1, no. 5: 221–223.

———. 1884. "Prevention and Restriction of Cholera." *Sanitarian* 13, no. 177: 107–113.

Sayre, L. A. 1870. "Minutes of the 63rd Annual Meeting." *Transactions of the Medical Society of the State of New York, 1870*, 6–74. Albany, NY: Van Benthuysen and Sons.

Select Committee on Medical Colleges and Societies of the New York Senate. 1846. "Report of the Committee on Medical Societies and Colleges, on Sundry Petitions from the Counties of Cayuga and Onondaga for a Homeopathic College." *Documents of the Senate of the State of New York* 4, no. 108: 1–4.

Seymour, William P. 1857. "History of the Cholera Epidemic of 1854, at Troy, NY." *Transactions of the Medical Society of the State of New York*, 175–202. Albany, NY: Van Benthuysen and Sons.

Shakespeare, E. O. 1887. "Address on Some New Aspects of the Cholera Question since the Discovery by Koch of the Comma Bacillus." *Journal of the American Medical Association* 8: 477–484.

———. 1890. *Report on Cholera in Europe and India*. Washington, DC: Government Printing Office.

Sharp, William. 1856. *The Advantages of Homeopathy*. New York: William Radde.

Shattuck, Lemuel. 1850. *Report of the Sanitary Commission of Massachusetts*. Boston: Dutton and Wentworth, State Printers.

Smith, Stephen. 1869. "Report on the Measures of Prevention and Relief to Be Adopted during the Prevalence of Epidemic Cholera." *Bulletin of the New York Academy of Medicine*, vol. 3, 59–72. New York: William Wood.

———. 1911. *The City That Was*. New York: Frank Alaben.

Spencer, C. L. 1857. *Lecture on the Philosophy and Claims of Homeopathy*. New York: H. Ludwig.

Spencer, Thomas. 1833. "Annual Address on the Nature of Epidemic Cholera, Usually Called Asiatic Cholera." *Transactions of the Medical Society of the State of New York, 1832, 1833*, 217–341. Albany, NY: Webster and Skinners.

Squibb, E. R. 1877. "Inaugural Address." *Transactions of the Medical Society of the State of New York, 1877*, 6. Albany, NY: Van Benthuysen and Sons.

Sterling, John W. 1849. "History of the Asiatic Cholera at Quarantine, Staten Island, New York in December 1848 and January 1849." *New York Journal of Medicine* 2, no. 3: 9–29.

Stow, T. Dwight. 1864. "Homeopathy in the Army. Medical Bigotry," *Transactions of the Homeopathic Medical Society of the State of New York for the Year, 1864*, 246–260. Albany, NY: The Argus Company, Printers.

Thayer, William Sydney.1969. *Osler and Other Papers*. Freeport, NY: Books for Libraries Press.

Thomson, John. 1841. "Dr. John Thomson's Petition in Behalf of the Thomsonians." *Boston Thomsonian Manual* 7, no. 9: 170–173.

Thomson, Samuel. 1822. *New Guide to Health*. Boston: E. G. House.

———. 1825. *Narrative of the Life and Medical Discoveries of Samuel Thomson*. Boston: E. G. House.

———. 1839. *The Law of Libel: Report of the Trial of Dr. Samuel Thomson, for an Alleged Libel in Warning the Public against the Impositions of Paine D. Badger*. Boston: Printed by Henry P. Lewis.

Thomsonian Botanic Watchman. 1834a . "New York Delusions," *Thomsonian Botanic Watchman* 1, no. 9: 129–132.

———. 1834b. "Medical Legislation." *Thomsonian Botanic Watchman* 1, no. 6: 86.

Thomsonian Messenger. 1843. "Understanding the Human System." *Thomsonian Messenger* 2, no. 10: 74.

Tooker, Robert N. 1885. *Homeopathy and its Relation to the Germ Theory*. Chicago: Gross and Delbridge.

Twain, Mark. 2010. *Autobiography of Mark Twain: The Complete and Authoritative Edition*. Berkeley: University of California Press.

Weiss, Ernst. 2009. *Georg Letham: Physician and Murderer*. Translated by Joel Rotenberg. New York: Archipelago Books.

Welch, William. 1893. "Asiatic Cholera in its Relations to Sanitary Reforms." Reprinted in *Popular Health Magazine*. Washington, DC.

———. 1920a. *Papers and Addresses by William Henry Welch*. Vol. 2, *Bacteriology*. Baltimore, MD: Johns Hopkins University Press.

———. 1920b. *Papers and Addresses by William Henry Welch*. Vol. 3, *Medical Education, History, and Miscellaneous*. Baltimore, MD: Johns Hopkins University Press.

Wells, P. P. 1864. "The Microscope in Pathological Investigations." *Transactions of the Homeopathic Medical Society of the State of New York for the Year 1864*, 90–92. Albany, NY: The Argus Company, Printers.

Whitney, Caspar. September 24, 1892. "Two Days with the Cholera Exiles." *Harper's Weekly*: 919–921.

Whitney, Daniel H. 1833. *The Family Physician and Guide to Health*. Philadelphia: H. Gilbert.

Wilcox, Dewitt G. 1904. *The Future of Homeopathy*. Rochester, NY: Democrat and Chronicle.

Williams, C. B. 1844. "Dr. C. J. B. Williams on Bilious Cholera." *Boston Medical and Surgical Journal* 31, no. 20: 393–394.

Wood, James C. 1899. "Relations of Homeopathy to Allied Systems of Therapeutics." *Transactions of the American Institute of Homeopathy*, vol. 52, 107–117. New York: Rooney and Otten Printing.

———. 1902. "Presidential Address." *Transactions of the American Institute of Homeopathy*, vol. 55, 34–50. Chicago: Publication Committee.

Yates, Christopher C. 1832. *Observations on the Epidemic Now Prevailing in the City of New York, Called the Asiatic or Spasmodic Cholera*. New York: Collins and Hannay.

SECONDARY SOURCES

Abbott, Andrew. 1988. *The System of Professions: An Essay on the Division of Expert Labor*. Chicago: University of Chicago Press.

———. 2005. "Linked Ecologies: States and Universities as Environments for Professions." *Sociological Theory* 23, no. 3: 245–274.

Abend, Gabriel. 2006. "Styles of Sociological Thought: Sociologies, Epistemologies, and the Mexican and US Quests for Truth." *Sociological Theory* 24, no. 1: 1–41.

Ackerknecht, Edward H. 1967. *Medicine at the Paris Hospital, 1794–1848*. Baltimore, MD: Johns Hopkins University Press.

Adams, George Worthington. 1996. *Doctors in Blue: The Medical History of the Union Army in the Civil War*. Baton Rouge: Louisiana State University Press.

Alcabes, Philip. 2009. *Dread: How Fear and Fantasy Have Fueled Epidemics from the Black Death to Avian Flu*. New York: Public Affairs.

Altman, Lawrence K. 1998. *Who Goes First? The Story of Self-Experimentation in Medicine*. Berkeley: University of California Press.

Baker, Jean H. 1983. *Affairs of Party: The Political Culture of Northern Democrats in the Mid-Nineteenth Century*: New York: Fordham University Press.

Baker, Samuel L. 1984. "Physician Licensure Laws in the United-States, 1865–1915." *Journal of the History of Medicine and Allied Sciences* 39, no. 2: 173–197.

Banta, H. D. 1971. "Flexner, a Reappraisal." *Social Science and Medicine* 5, no. 6: 655–661.

Barker, Kristin. 2005. *The Fibromyalgia Story: Medical Authority and Women's Worlds of Pain*. Philadelphia: Temple University Press.

Barnes, David S. 1995. *The Making of a Social Disease: Tuberculosis in Nineteenth-Century France*. Berkeley: University of California Press.

Barrett, Frank A. 1996. "Daniel Drake's Medical Geography." *Social Science and Medicine* 42, no. 6: 791–800.

Barry, John. 2005. *The Great Influenza: The Epic Story of the Deadliest Plague in History*. New York: Penguin.

Ben-David, Joseph. 1960. "Scientific Productivity and Academic Organization in Nineteenth Century Medicine." *American Sociological Review* 25, no. 6: 828–843.

Bender, Thomas. 1976. "Science and the Culture of American Communities: The Nineteenth Century." *History of Education Quarterly* 16, no. 1: 63–77.

Benford, Robert D., and David A. Snow. 2000. "Framing Processes and Social Movements: An Overview and Assessment." *Annual Review of Sociology* 26: 611–639.

Benson, Lee. 1961. *The Concept of Jacksonian Democracy; New York as a Test Case*. Princeton, NJ: Princeton University Press.

Berlant, Jeffrey Lionel. 1975. *Profession and Monopoly: A Study of Medicine in the United States and Great Britain*. Berkeley: University of California Press.

Berliner, Howard S. 1985. *A System of Scientific Medicine: Philanthropic Foundations in the Flexner Era*. New York: Tavistock.

Berman, Alex, and Michael A. Flannery. 2001. *America's Botanico-Medical Movements*. New York: Pharmaceutical Products Press.

Biagioli, Mario. 1994. *Galileo, Courtier: The Practice of Science in the Culture of Absolutism*. Chicago: University of Chicago Press.

Bilson, Geoffrey. 1980. *A Darkened House: Cholera in Nineteenth-Century Canada*. Toronto: University of Toronto Press.

Birenbaum, Arnold. 2002. *Wounded Profession: American Medicine Enters the Age of Managed Care*. Westport, CT: Praeger.

Bloor, David. 1991. *Knowledge and Social Imagery*. Chicago: University of Chicago Press.

Boggs, S. W. 1947. "Cartohypnosis." *Scientific Monthly* 64: 469–476.

BonJour, Laurence. 1978. "Can Empirical Knowledge Have a Foundation?" *American Philosophical Quarterly* 15, no. 1: 1–13.

Bonner, Thomas Neville. 1963. *American Doctors and German Universities*. Lincoln: University of Nebraska Press.

Bordley, James, and Abner McGehee Harvey. 1976. *Two Centuries of American Medicine, 1776–1976*. Philadelphia: WB Saunders.

Bornside, George H. 1982. "Waldemar Haffkine's Cholera Vaccines and the Ferran-Haffkine Priority Dispute." *Journal of the History of Medicine and Allied Sciences* 37, no. 4: 399.

Borrell, Merriley. 1987. "Instrumentation and the Rise of Modern Physiology." *Science and Technology Studies* 5, no. 2: 53–62.

Brannigan, Augustine. 1981. *The Social Basis of Scientific Discoveries*. New York: Cambridge University Press.

Briggs, Asa. 1961. "Cholera and Society in the Nineteenth Century." *Past and Present* 19, no. 1: 76–96.

Briggs, Charles L., and Clara Mantini-Briggs. 2003. *Stories in the Time of Cholera: Racial Profiling during a Medical Nightmare*. Berkeley: University of California Press.

Brock, Thomas D. 1988. *Robert Koch, a Life in Medicine and Bacteriology*. Madison, WI: Science Tech Publishers.

Brown, E. Richard. 1979. *Rockefeller Medicine Men: Medicine and Capitalism in America*. Berkeley: University of California Press.

Bruner, Jerome. 1991. "The Narrative Construction of Reality." *Critical Inquiry* 18, no. 1: 1–21.

Bulloch, William. 1979. *The History of Bacteriology*. New York: Dover Publications.

Burrell, Sean, and Geoffrey Gill. 2005. "The Liverpool Cholera Epidemic of 1832 and Anatomical Dissection—Medical Mistrust and Civil Unrest." *Journal of the History of Medicine and Allied Sciences* 60, no. 4: 478–498.

Burrow, James Gordon. 1963. *AMA: Voice of American Medicine*. Baltimore, MD: Johns Hopkins University Press.

Callon, Michel. 1986. "Some Elements of a Sociology of Translation: Domestication of the Scallops and the Fishermen of St Brieuc Bay." In *Power, Action and Belief: A New Sociology of Knowledge*, edited by John Law, 196–233. London: Routledge and Kegan Paul.

Carruthers, Bruce G., and Wendy Espeland. 1991. "Accounting for Rationality—Double-Entry Bookkeeping and the Rhetoric of Economic Rationality." *American Journal of Sociology* 97, no. 1: 31–69.

Cassedy, John H. 1962. *Charles V. Chapin and the Public Health Movement*. Cambridge, MA: Harvard University Press.

———. 1984. *American Medicine and Statistical Thinking, 1800–1860*. Cambridge, MA: Harvard University Press.

Chambers, John Sharpe. 1938. *The Conquest of Cholera: America's Greatest Scourge*. New York: Macmillan.

Chase, Marilyn. 200. *The Barbary Plague: The Black Death in Victorian San Francisco*. New York: Random House.

Chernow, Ron. 1998. *Titan: The Life of John D. Rockefeller, Sr*. New York: Random House.

Clarke, Adele, Janet Shim, Laura Mamo, Jennifer Ruth Fosket, and Jennifer R. Fishman. 2003. "Biomedicalization: Technoscientific Transformations of Health, Illness, and US Biomedicine." *American Sociological Review* 68, no. 2: 161–194.

Code, Lorraine. 1995. *Rhetorical Spaces: Essays on Gendered Locations*. New York: Routledge.

Cohen, Patricia Cline. 1982. *A Calculating People: The Spread of Numeracy in Early America*. Chicago: University of Chicago Press.

Coleman, William. 1987. "Koch's Comma Bacillus: The First Year." *Bulletin of the History of Medicine* 61, no. 3: 315–342.

Collins, Harry. 1992. *Changing Order: Replication and Induction in Scientific Practice*. Chicago: University of Chicago Press.

Collins, Randall. 2000. *The Sociology of Philosophies: A Global Theory of Intellectual Change*. Cambridge, MA: Harvard University Press.

Colwell, Rita R. 2002. "A Voyage of Discovery: Cholera, Climate and Complexity." *Environmental Microbiology* 4, no. 2: 67–69.

Colwell, Rita R., and Anwar Huq. 2001. "Marine Ecosystems and Cholera." *Hydrobiologia* 460, no. 1–3: 141–145.

Corner, George Washington. 1964. *A History of the Rockefeller Institute, 1901–1953*. New York: Rockefeller Institute Press.

Coulter, Harris L. 1969. "Political and Social Aspects of Nineteenth-Century Medicine in the United States: The Formation of the American Medical Association and Its Struggle with Homeopathic and Eclectic Physicians." PhD diss., Columbia University.

———. 1973. *Divided Legacy: A History of the Schism in Medical Thought*. Washington, DC: Wehawken Book Co.

Crane, Diana. 1972. *Invisible Colleges: Diffusion of Knowledge in Scientific Communities*. Chicago: University of Chicago Press.

Crosby, Molly Caldwell. 2007. *The American Plague: The Untold Story of Yellow Fever, the Epidemic That Shaped Our History*. New York: Berkley Books.

Cunningham, Andrew. 2002. "Transforming Plague: The Laboratory and the Identity of Infectious Disease." In *The Laboratory Revolution in Medicine*, edited by Andrew Cunningham and Perry Williams, 209–224. New York: Cambridge University Press.

Daston, Lorraine. 1992. "Objectivity and the Escape from Perspective." *Social Studies of Science* 22, no. 4: 597–618.

Daston, Lorraine, and Peter Galison. 2010. *Objectivity*. New York: Zone Books.

Davidson, Arnold I. 2001. *The Emergence of Sexuality: Historical Epistemology and the Formation of Concepts*. Cambridge, MA: Harvard University Press.

Davis, Joseph. E. 2002. *Stories of Change: Narrative and Social Movements*. Albany: State University of New York Press.

Dear, Peter. 1992. "From Truth to Disinterestedness in the 17th-Century." *Social Studies of Science* 22, no. 4: 619–631.

DeGloma, Thomas. 2010. "Awakenings: Autobiography, Memory, and the Social Logic of Personal Discovery." *Sociological Forum* 25, no. 3: 519–540.

de Kruif, Paul. 1996. *Microbe Hunters*. San Diego, CA: Harcourt Brace.

De Ville, Kenneth. A, and R. B. Freeman, eds. 1990. *Medical Malpractice in Nineteenth-Century America*. New York: New York University Press.

DiMaggio, Paul J., and Walter W. Powell. 1983. "The Iron Cage Revisited—Institutional Isomorphism and Collective Rationality in Organizational Fields." *American Sociological Review* 48, no. 2: 147–160.

Diner, Steven J. 1998. *A Very Different Age: Americans of the Progressive Era*. New York: Hill and Wang.

Douglas, Mary. 1986. *How Institutions Think*. Syracuse, NY: Syracuse University Press.

Downer, John. 2011. " '737-Cabriolet': The Limits of Knowledge and the Sociology of Inevitable Failure." *American Journal of Sociology* 117, no. 3: 725–762.

Dubos, René. 1987. *Mirage of Health: Utopias, Progress and Biological Change*. New Brunswick, NJ: Rutgers University Press.

Duffy, John. 1968. *A History of Public Health in New York City: 1866–1966*. Vol. 1.New York: Russell Sage Foundation.

———. 1974. *A History of Public Health in New York City: 1866–1966*. Vol. 2. New York: Russell Sage Foundation.

———. 1990. *The Sanitarians: A History of American Public Health*. Urbana: University of Illinois Press.

———. 1993. *From Humors to Medical Science: A History of American Medicine*. Urbana: University of Illinois Press.

Dumit, Joseph. 2006. "Illnesses You Have to Fight to Get: Facts as Forces in Uncertain, Emergent Illnesses." *Social Science and Medicine* 62, no. 3: 577–590.

Durey, Michael. 1979. *The Return of the Plague: British Society and the Cholera, 1831–2*. New York: Gill and Macmillan.

Dzur, Albert W. 2004. "Civic Participation in Professional Domains." *The Good Society* 13, no. 1: 1–5.

Eisenberg, David M., Roger B. Davis, Susan L. Ettner, Scott Appel, Sonja Wilkey, Maria Van Rompay, and Ronald C. Kessler. 1998. "Trends in Alternative Medicine Use in the United States, 1990–1997: Results of a Follow-up National Survey." *JAMA* 280, no. 18: 1569–1575.

Ellis, Jack D. 1990. *The Physician-Legislators of France: Medicine and Politics in the Early Third Republic, 1870–1914*. New York: Cambridge University Press.

Ellis, John H. 1992. *Yellow Fever and Public Health in the New South*. Lexington: University Press of Kentucky.

Epstein, Steven. 1996. *Impure Science: AIDS, Activism, and the Politics of Knowledge*. Berkeley: University of California Press.

Evans, Richard J. 2005. *Death in Hamburg: Society and Politics in the Cholera Years*. New York: Penguin Books.

Ewald, Paul W. 2002. *Plague Time: The New Germ Theory of Disease*. New York: Anchor Books.

Eyerman, Ron, and Andrew Jamison. 1991. *Social Movements: A Cognitive Approach*. University Park: Pennsylvania State University Press.

Eyler, John M. 1973. "William Farr on the Cholera: The Sanitarian's Disease Theory and the Statistician's Method." *Journal of the History of Medicine and Allied Sciences* 28, no. 2: 79–100.

Fangerau, H. M. 2006. "The Novel Arrowsmith, Paul de Kruif (1890–1971) and Jacques Loeb (1859–1924): A Literary Portrait of 'Medical Science.'" *Medical Humanities* 32, no. 2: 82–87.

Farmer, Paul. 2001. *Infections and Inequalities: The Modern Plagues*. Berkeley: University of California Press.

Faust, Drew G. 2009. *This Republic of Suffering: Death and the American Civil War*. New York: Vintage.

Fee, Elizabeth, and Evelynn M. Hammonds. 1995. "Science, Politics and the Art of Persuasion." In *Hives of Sickness: Public Health and Epidemics in New York City*, edited by David Rosner, 155—196. New Brunswick, NJ: Rutgers University Press.

Feller, Daniel 1990. "Politics and Society: Toward a Jacksonian Synthesis." *Journal of the Early Republic* 10, no. 2: 135–161.

Fishbein, Morris 1947. *A History of the American Medical Association, 1847 to 1947*. Philadelphia: Saunders.

Fleck, Ludwig. 1979. *Genesis and Development of a Scientific Fact*. Chicago: University of Chicago Press.

Fleming, Donald. 1987. *William H. Welch and the Rise of Modern Medicine*. Baltimore, MD: Johns Hopkins University Press.

Fortun, Michael A. 2008. *Promising Genomics: Iceland and deCODE Genetics in a World of Speculation*. Berkeley: University of California Press.

Foucault, Michel. 1980. *Power/Knowledge: Selected Interviews and Other Writings, 1972–1977*. Brighton, UK: Harvester Press.

———. 1994. *The Birth of the Clinic: An Archaeology of Medical Perception*. New York: Vintage.

———. 2002. *Archaeology of Knowledge* New York: Routledge.

Fourcade, Marion. 2009. *Economists and Societies: Discipline and Profession in the United States, Britain, and France, 1890s to 1990s*. Princeton, NJ: Princeton University Press.

Fourcade-Gourinchas, Marion. 2001. "Politics, Institutional Structures, and the Rise of Economics: A Comparative Study." *Theory and Society* 30, no. 3: 397–447.

Freidson, Eliot. 1970. *Professional Dominance: The Social Structure of Medical Care*. New York: Atherton Press.

———. 1986. *Professional Powers: A Study of the Institutionalization of Formal Knowledge*. Chicago: University of Chicago Press.

———. 1988. *Profession of Medicine: A Study of the Sociology of Applied Knowledge*. Chicago: University of Chicago Press.

———. 2001. *Professionalism: The Third Logic*. Chicago: University of Chicago Press.

Frickel, Scott. 2004. *Chemical Consequences: Environmental Mutagens, Scientist Activism, and the Rise of Genetic Toxicology*. New Brunswick, NJ: Rutgers University Press.

Frickel, Scott, and Neil Gross. 2005. "A General Theory of Scientific/Intellectual Movements." *American Sociological Review* 70, no. 2: 204–232.

Fricker, Miranda. 2007. *Epistemic Injustice: Power and the Ethics of Knowing*. New York: Oxford University Press.

Frieden, Nancy M. 1977. "The Russian Cholera Epidemic, 1892–93, and Medical Professionalization." *Journal of Social History* 10, no. 4: 538–559.

Fuchs, Stephan. 1992. *The Professional Quest for Truth: A Social Theory of Science and Knowledge*. Albany: State University of New York Press.

Fuller, Steve. 2002. *Social Epistemology*. Bloomington: Indiana University Press.

Geison, Gerald L. 1984. *Professions and the French State, 1700–1900*. Philadelphia: University of Pennsylvania Press.

George, Alexander L., and Andrew Bennett. 2005. *Case Studies and Theory Development in the Social Sciences*. Cambridge, MA: The MIT Press.

Gevitz, Norman. 1992. "The Fate of Sectarian Medical Education." In *Beyond Flexner: Medical Education in the Twentieth Century*, edited by Norman Barzansky, 83–97.New York: Greenwood Press.

Giddens, Anthony. 1984. *The Constitution of Society: Outline of the Theory of Structuration*. Berkeley: University of California Press.

Gieryn, Thomas. 1983. "Boundary-Work and the Demarcation of Science from Non-Science: Strains and Interests in Professional Ideologies of Scientists." *American Sociological Review* 48, no. 6: 781–795.

———. 1999. *Cultural Boundaries of Science: Credibility on the Line*. Chicago: University of Chicago Press.

Gilbert, E. W. 1958. "Pioneer Maps of Health and Disease in England." *Geographical Journal* 124, no. 2: 172–183.

Gilbert, G. Nigel, and Michael Mulkay. 1984. *Opening Pandora's Box: A Sociological Analysis of Scientists' Discourse*. New York: Cambridge University Press.

Gilbert, Pamela K. 2008. *Cholera and Nation: Doctoring the Social Body in Victorian England*. Albany: State University of New York Press.

Ginzburg, Carlo. 1980. "Morelli, Freud and Sherlock Holmes: Clues and Scientific Method." *History Workshop*, no. 9: 5–36.

———. 1992. *The Cheese and the Worms: The Cosmos of a Sixteenth-Century Miller*. Baltimore, MD: Johns Hopkins University Press.

Glaeser, Andreas. 2011. *Political Epistemics: The Secret Police, the Opposition, and the End of East German Socialism*. Chicago: University of Chicago Press.

Goldstein, Jan E. 1990. *Console and Classify: The French Psychiatric Profession in the Nineteenth Century*. New York: Cambridge University Press.

Gossel, Patricia Peck. 1992. "A Need for Standard Methods: The Case of American Bacteriology." In *The Right Tools for the Job*, edited by Adele E. Clarke and Joan H. Fujimora, 287–311. Princeton, NJ: Princeton University Press.

Gradmann, Christoph. 2008. "A Matter of Methods: The Historicity of Koch's Postulates 1840–2000." *Medizinhistorisches Journal* 43, no. 2: 121–148.

———. 2009. *Laboratory Disease: Robert Koch's Medical Bacteriology*. Baltimore, MD: Johns Hopkins University Press.

Grob, Gerald N. 2002. *The Deadly Truth: A History of Disease in America*. Cambridge, MA: Harvard University Press.

Guston, David H. 1999. "Stabilizing the Boundary between US Politics and Science: The Role of the Office of Technology Transfer as a Boundary Organization." *Social Studies of Science* 29, no. 1: 87–111.

Hacking, Ian. 1983. *Representing and Intervening: Introductory Topics in the Philosophy of Natural Science*. New York: Cambridge University Press.

———. 1985. "Styles of Scientific Reasoning." In *Post-analytic Philosophy*, edited by John Rajchman and Cornel West, 145–164. New York: Columbia University Press.

———. 1990. *The Taming of Chance*. New York: Cambridge University Press.

Haller, John S. 1981. *American Medicine in Transition 1840–1910*. Urbana: University of Illinois Press.

———. 2000. *The People's Doctors: Samuel Thomson and the American Botanical Movement, 1790–1860*. Carbondale: Southern Illinois University Press.

———. 2005. *The History of American Homeopathy: The Academic Years, 1820–1935*. Binghamton, NY: Informa HealthCare.

Halttunen, Karen. 1986. *Confidence Men and Painted Women: A Study of Middle-Class Culture in America, 1830–1870*. New Haven, CT: Yale University Press.

Hamlin, Christopher. 2009. *Cholera: The Biography*. New York: Oxford University Press.

Hammonds, Evelynn. 1999. *Childhood's Deadly Scourge: The Campaign to Control Diphtheria in New York City, 1880–1930*. Baltimore, MD: Johns Hopkins University Press.

Hansen, Bert. 1999. "New Images of a New Medicine: Visual Evidence for the Widespread Popularity of Therapeutic Discoveries in America after 1885." *Bulletin of the History of Medicine* 73, no. 4: 629–678.

Haraway, Donna. 2006. *Primate Visions: Gender, Race, and Nature in the World of Modern Science*. New York: Routledge.

Harding, Sandra G. 1986. *The Science Question in Feminism*. Ithaca, NY: Cornell University Press.

———. 1998. *Is Science Multicultural? Postcolonialisms, Feminisms, and Epistemologies*. Bloomington: Indiana University Press.

Haskell, Thomas L. 1985. "Capitalism and the Origins of the Humanitarian Sensibility, Part 1." *American Historical Review* 90, no. 2: 339–361.

Hess, David J. 2004. "Medical Modernisation, Scientific Research Fields and the Epistemic Politics of Health Social Movements." *Sociology of Health and Illness* 26, no. 6: 695–709.

Hilgartner, Stephen. 2000. *Science on Stage: Expert Advice as Public Drama*. Stanford, CA: Stanford University Press.

Hofstadter, Richard. 1963. *Anti-Intellectualism in American Life*. New York: Knopf.

Howe, Daniel Walker. 2007. *What Hath God Wrought: The Transformation of America, 1815–1848*. New York: Oxford University Press.

Humphreys, Margaret. 1999. *Yellow Fever and the South*. Baltimore, MD: Johns Hopkins University Press.

———. 2002. "No Safe Place: Disease and Panic in American History." *American Literary History* 14, no. 4: 845–857.

James, William. 1890. *The Principles of Psychology*: New York: Henry Holt.

———. 1909. *A Pluralistic Universe*. New York: Longmans, Green.

———. 1976. *Essays in Radical Empiricism*. Cambridge, MA: Harvard University Press.

Jarcho, Saul. 1970. "Yellow Fever, Cholera, and the Beginnings of Medical Cartography." *Journal of the History of Medicine and Allied Sciences* 25, no. 2: 131–142.

Jardine, Nicholas. 1992. "The Laboratory Revolution in Medicine as Rhetorical and Aesthetic Accomplishment.'" In *The Laboratory Revolution in Medicine*, edited by Andrew Cunningham and Perry Williams, 304–323. New York: Cambridge University Press.

Jasper, James M. 2006. *Getting Your Way: Strategic Dilemmas in the Real World*. Chicago: University of Chicago Press.

Jewson, N. D. 1974. "Medical Knowledge and Patronage System in 18th Century England." *Sociology* 8, no. 3: 269–385.

———. 1976. "The Disappearance of the Sick-Man from Medical Cosmology 1870–1970." *Sociology* 10, no. 2: 225–44.

Johnson, Steven. 2006. *The Ghost Map: The Story of London's Most Terrifying Epidemic—and How It Changed Science, Cities, and the Modern World*. New York: Riverhead Books.

Jones, Russell M. 1970. "American Doctors in Paris, 1820–1861: A Statistical Profile." *Journal of the History of Medicine and Allied Sciences* 25, no. 2: 143–157.

Jonsen, Albert, and Stephen Toulmin. 1988. *The Abuse of Casuistry: A History of Moral Reasoning*. Berkeley: University of California Press.

Kass, Alvin. 1965. *Politics in New York State, 1800–1830*. Syracuse, NY: Syracuse University Press.

Kater, Michael. 1985. "Professionalization and Socialization of Physicians in Wilhelmine and Weimar Germany." *Journal of Contemporary History* 20, no. 4: 677–701.

Katz, Jay. 2002. *The Silent World of Doctor and Patient*. Baltimore, MD: Johns Hopkins University Press.

Kaufman, Martin. 1988. "Homeopathy in America: The Rise and Fall and Persistence of a Medical Heresy." In *Other Healers: Unorthodox Medicine in America*, edited by Norman Gevitz, 99–123. Ann Arbor: University of Michigan Press.

Kelman, Steven. 1987. "The Political Foundations of American Statistical Policy." In *The Politics of Statistics*, edited by William Alonso and Paul Starr, 275–302. New York: Russell Sage Foundation.

Kett, Joseph F. 1968. *The Formation of the American Medical Profession: The Role of Institutions, 1780–1860*. New Haven, CT: Yale University Press.

Kim, Jaegwon. 1988. "What Is Naturalized Epistemology?" *Philosophical Perspectives*, no. 2: 381–405.

Kleinman, Daniel Lee, and Abby J. Kinchy. 2003. "Boundaries in Science Policy Making: Bovine Growth Hormone in the European Union." *Sociological Quarterly* 44, no. 4: 577–595.

Knorr-Cetina, Karin. 1999. *Epistemic Cultures: How the Sciences Make Knowledge*. Cambridge, MA: Harvard University Press.

Koch, Tom. 2005. *Cartographies of Disease: Maps, Mapping, and Medicine*. Redlands, CA: ESRI Press.

Kohl, Lawrence Frederick. 1989. *The Politics of Individualism: Parties and the American Character in the Jacksonian Era*. New York: Oxford University Press.

Kohler, Robert E. 2002. *Landscapes and Labscapes: Exploring the Lab-Field Border in Biology*. Chicago: University of Chicago Press.

Krause, Elliott A. 1996. *Death of the Guilds: Professions, States, and the Advance of Capitalism, 1930 to the Present*. New Haven, CT: Yale University Press.

Kraut, Alan M. 1995. *Silent Travelers: Germs, Genes, and the "Immigrant Menace."* Baltimore, MD: Johns Hopkins University Press.

Kroll, Steven, Steve Kroll-Smith, and H. Hugh Floyd. 2000. *Bodies in Protest: Environmental Illness and the Struggle over Medical Knowledge*. New York: New York University Press.

Kudlick, Catherine J. 1996. *Cholera in Post-Revolutionary Paris: A Cultural History*. Berkeley: University of California Press.

Kuhn, Thomas S. 1962. "Historical Structure of Scientific Discovery." *Science* 136, no. 1358: 760–764.

———. 1996. *The Structure of Scientific Revolutions*. Chicago: University of Chicago Press.

Kurzman, Charles. 1994. "Epistemology and the Sociology of Knowledge." *Philosophy of the Social Sciences* 24, no. 3: 267–290.

Lachmund, Jens. 1998. "Between Scrutiny and Treatment: Physical Diagnosis and the Restructuring of 19th Century Medical Practice." *Sociology of Health and Illness* 20, no. 6: 779–801.

Lakoff, Andrew. 2005. *Pharmaceutical Reason: Knowledge and Value in Global Psychiatry*. New York: Cambridge University Press.

Larson, Magali Sarfatti. 1977. *The Rise of Professionalism: A Sociological Analysis*. Berkeley: University of California Press.

Latour, Bruno. 1987. *Science in Action: How to Follow Scientists and Engineers through Society*. Cambridge, MA: Harvard University Press.

———. 1988. *The Pasteurization of France*. Cambridge, MA: Harvard University Press.

———. 2005. *Reassembling the Social: An Introduction to Actor-Network-Theory*. New York: Oxford University Press.

Latour, Bruno, and Steve Woolgar. 1986. *Laboratory Life: The Construction of Scientific Facts*. Princeton, NJ: Princeton University Press.

Law, John. 1992. "Notes on the Theory of the Actor-Network: Ordering, Strategy, and Heterogeneity." *Systemic Practice and Action Research* 5, no. 4: 379–393.

Leavitt, Judith Walzer. 1992. " 'Typhoid Mary' Strikes Back: Bacteriological Theory and Practice in Early Twentieth-Century Public Health." *Isis* 83, no. 4: 608–629.

Lewis, Michael. 2004. *Moneyball: The Art of Winning an Unfair Game*. New York: W. W. Norton.

Lewis, Sinclair. 2008. *Arrowsmith*. New York: Signet Classics.

Light, Donald, and Sol Levine. 1988. "The Changing Character of the Medical Profession: A Theoretical Overview." *Milbank Quarterly* 66, suppl. 2: 10–32.

Longino, Helen. E. 2002. *The Fate of Knowledge*. Princeton, NJ: Princeton University Press.

Longmate, Norman. 1966. *King Cholera: The Biography of a Disease*: London: Hamilton.

Löwy, Ilana. 1992. "From Guinea Pigs to Man: The Development of Haffkine's Anticholera Vaccine." *Journal of the History of Medicine and Allied Sciences* 47, no. 3: 270–309.

Ludmerer, Kenneth. 1985. *Learning to Heal*. New York: Basic Books.

Mallard, Gregoire, Michele Lamont, and Joshua Guetzkow. 2009. "Fairness as Appropriateness." *Science, Technology and Human Values* 34, no. 5: 573–606.

Mannheim, Karl. 1992. *Essays on the Sociology of Culture*. New York: Routledge.

March, James, and Simon, Herbert. 1958. *Organizations*. New York: Wiley.

Marcus, Alan. 1979. "Disease Prevention in America: From a Local to a National Outlook, 1880–1910." *Bulletin of the History of Medicine* 53, no. 2: 184–203.

Markel, Howard. 1997. *Quarantine! East European Jewish Immigrants and the New York City Epidemics of 1892*. Baltimore, MD: Johns Hopkins University Press.

———. 2008. The Principles and Practice of Medicine." *Journal of the American Medical Association* 299, no. 10: 1199.

Markel, Howard, and Alexandra Minna Stern. 2002. "The Foreignness of Germs: The Persistent Association of Immigrants and Disease in American Society." *Milbank Quarterly* 80, no. 4: 757–788.

Markowitz, Gerald E., and David Rosner. 1973. "Doctors in Crisis: A Study of the Use of Medical Education Reform to Establish Modern Professional Elitism in Medicine." *American Quarterly* 25, no. 1: 83–107.

Marks, Geoffrey, and William K. Beatty. 1973. *The Story of Medicine in America*. New York: Scribner.

Marlor, Chantelle. 2010. "Bureaucracy, Democracy and Exclusion: Why Indigenous Knowledge Holders Have a Hard Time Being Taken Seriously." *Qualitative Sociology* 33, no. 4: 1–19.

Matthews, J. Rosser. 1995. *Quantification and the Quest for Medical Certainty*. Princeton, NJ: Princeton University Press.

Maulitz, Russell C. 1979. "Physician Versus Bacteriologist: The Ideology of Science in Clinical Medicine." In *The therapeutic Revolution: Essays in the Social History of American Medicine*, edited by Morris J. Vogel and Charles Rosenberg, 91–107. Philadelphia: University of Pennsylvania Press.

———. 1982. "Robert Koch and American-Medicine." *Annals of Internal Medicine* 97, no. 5: 761–766.

McClary, Andrew. 1980. "Germs Are Everywhere: The Germ Threat as Seen in Magazine Articles." *Journal of American Culture* 3, no. 1: 33–46.

McCloskey, Deidre N. 1985. *The Rhetoric of Economics*. Madison: University of Wisconsin Press.

McKeown, Thomas. 1976. *The Modern Rise of Population*. London: Edward Arnold.

———. 1979. *The Role of Medicine: Dream, Mirage, or Nemesis?* Princeton, NJ: Princeton University Press.

Meckel, Richard A. 1998. *Save the Babies: American Public Health Reform and the Prevention of Infant Mortality, 1850–1929*. Ann Arbor: University of Michigan Press.

Menand, Louis. 2001. *The Metaphysical Club*. New York: Macmillan.

Merton, Robert K. 1968. *Social Theory and Social Structure*. New York: Free Press.

———. 1973. *The Sociology of Science: Theoretical and Empirical Investigations*. Chicago: University of Chicago Press.

Meyers, Marvin. 1957. *The Jacksonian Persuasion: Politics and Belief*. Stanford, CA: Stanford University Press.

Mindich, David. 2000. *Just the Facts: How "Objectivity" Came to Define American Journalism*. New York: New York University Press.

Mitman, Gregg, and Ronald L. Numbers. 2003. "From Miasma to Asthma: The Changing Fortunes of Medical Geography in America." *History and Philosophy of the Life Sciences* 25, no. 3: 391–412.

Montgomery, Kathryn. 2006. *How Doctors Think: Clinical Judgment and the Practice of Medicine*. New York: Oxford University Press.

Moore, Kelley. 1996. "Organizing Integrity: American Science and the Creation of Public Interest Organizations, 1955–1975." *American Journal of Sociology* 101, no. 6: 1592–1627.

Morris, Robert John. 1976. *Cholera 1832: The Social Response to an Epidemic*. London: Holm and Meier.

Murphy, Claudette Michelle. 2006. *Sick Building Syndrome and the Problem of Uncertainty: Environmental Politics, Technoscience, and Women Workers*. Durham, NC: Duke University Press.

Nagel, Thomas. 1989. *The View from Nowhere*. New York: Oxford University Press.

Novick, Peter. 1988. *That Noble Dream: The "Objectivity Question" and the American Historical Profession*. New York: Cambridge University Press.

Numbers, Ronald L. 1978. *Almost Persuaded: American Physicians and Compulsory Health Insurance, 1912–1920*. Baltimore, MD: Johns Hopkins University Press.

———. 1988. "The Fall and Rise of the American Medical Profession." In *The Professions in American History*, edited by Nathan O. Hatch, 51–72. Notre Dame, IN: University of Notre Dame Press.

Ogawa, M. 2000. "Uneasy Bedfellows: Science and Politics in the Refutation of Koch's Bacterial Theory of Cholera." *Bulletin of the History of Medicine* 74, no. 4: 671–707.

Oleson, Alexandra, and Voss, John. 1979. *The Organization of Knowledge in Modern America, 1860–1920*. Baltimore, MD: Johns Hopkins University Press.

Osborne, Michael A. 2000. "The Geographical Imperative in Nineteenth-Century French Medicine." *Medical History Supplement* 20: 31–50.

Parsons, Talcott. 1964. *Essays in Sociological Theory*. New York: Free Press.

———. 1991. *The Social System*. New York: Routledge.

Patterson, Wendy. 2002. *Strategic Narrative: New Perspectives on the Power of Personal and Cultural Stories*. Lanham, MD: Lexington Books.

Paul, Harry W. 1990. "Review of *The Pasteurization of France* by Bruno Latour." *American Journal of Sociology* 96, no. 1: 232–234.

Peckham, Howard Henry. 1994. *The Making of the University of Michigan, 1817–1992*. Ann Arbor: University of Michigan Bentley Library.
Peirce, Charles Saunders. 1955. *Philosophical Writings of Peirce*. New York: Dover.
———. 1978. *The Philosophy of Peirce: Selected Writings*. New York: AMS Press.
Penders, Bert, John Verbakel, and Annemeik Nelis. 2009. "The Social Study of Corporate Science: A Research Manifesto." *Bulletin of Science, Technology and Society* 29, no. 6: 439–446.
Perelman, Chaim, and Lucie Olbrechts-Tyteca. 1969. *The New Rhetoric: A Treatise on Argumentation*. Notre Dame, IN: University of Notre Dame Press.
Pessen, Edward. 1969. *Jacksonian America: Society, Personality, and Politics*. Homewood, IL: Dorsey Press.
Pickering, Andrew. 1984. *Constructing Quarks: A Sociological History of Particle Physics*. Chicago: University of Chicago Press.
Polanyi, Michael. 1958. *Personal Knowledge: Towards a Post-Critical Philosophy*. Chicago: University of Chicago Press.
Polletta, Francesca. 2006. *It Was Like a Fever: Storytelling in Protest and Politics*. Chicago: University of Chicago Press.
Poovey, Mary. 1998. *A History of the Modern Fact: Problems of Knowledge in the Sciences of Wealth and Society*. Chicago: University of Chicago Press.
Popper, Karl. 1992. *The Logic of Scientific Discovery*. New York: Routledge.
Porter, Roy. 1998. *The Greatest Benefit to Mankind: A Medical History of Humanity*. New York: W. W. Norton.
Porter, Theodore M. 1988. *The Rise of Statistical Thinking, 1820–1900*. Princeton, NJ: Princeton University Press.
———. 1994. "Making Things Quantitative." *Science in Context* 7, no. 3: 389–407.
———. 1995. *Trust in Numbers: The Pursuit of Objectivity in Science and Public Life*. Princeton, NJ: Princeton University Press.
Putnam, Hilary. 1995. *Words and Life*. Cambridge, MA: Harvard University Press.
———. 1999. *The Threefold Cord: Mind, Body, and World*. New York: Columbia University Press.
Quadagno, Jill. 2005. *One Nation, Uninsured: Why the US Has No National Health Insurance*. New York: Oxford University Press.
Reay, Mike. 2010. "Knowledge Distribution, Embodiment, and Insulation." *Sociological Theory* 28, no. 1: 91–107.
Reed, Isaac. 2011. *Interpretation and Social Knowledge: On the Use of Theory in the Human Sciences*. Chicago: University of Chicago Press.
Riessman, Catherine Kohler. 1993. *Narrative Analysis, Qualitative Research Methods Series*. London: Sage.
Reynolds, David. 2008. *Waking Giant: America in the Age of Jackson*. New York: Harper.
Richmond, Phyllis Allen. 1947. "Etiological Theory in America prior to the Civil War." *Journal of the History of Medicine and Allied Sciences* 2, no. 4: 489–520.
———. 1954. "American Attitudes toward the Germ Theory of Disease." *Journal of the History of Medicine and Allied Sciences* 9, no. 4: 428–454.

Romano, Terrie M. 1997. "The Cattle Plague of 1865 and the Reception of 'The Germ Theory' in Mid-Victorian Britain." *Journal of the History of Medicine and Allied Sciences* 52, no. 1: 51–80.

Rosen, George. 1993. *A History of Public Health*. Baltimore, MD: Johns Hopkins University Press.

Rosen, George, and Charles E. Rosenberg. 1983. *The Structure of American Medical Practice, 1875–1941*. Philadelphia: University of Pennsylvania Press.

Rosenberg, Charles E. 1963. "Martin Arrowsmith: The Scientist as Hero." *American Quarterly* 15, no. 3: 447–458.

———. 1966. "Cholera in Nineteenth-Century Europe—Tool for Social and Economic Analysis." *Comparative Studies in Society and History* 8, no. 4: 452–463.

———. 1977. "And Heal the Sick: The Hospital and the Patient in the 19th Century America." *Journal of Social History* 10, no. 4: 428–447.

———. 1987a. *The Care of Strangers: The Rise of America's Hospital System*. New York: Basic Books.

———. 1987b. *The Cholera Years: The United States in 1832, 1849, and 1866*. Chicago: University of Chicago Press.

———. 1995. *The Trial of the Assassin Guiteau: Psychiatry and the Law in the Gilded Age*. Chicago: University of Chicago Press.

———. 2002. "The Tyranny of Diagnosis: Specific Entities and Individual Experience." *Milbank Quarterly* 80, no. 2: 237–260.

Rosenkrantz, Barbara Guttmann. 1974. "Cart before Horse: Theory, Practice and Professional Image in American Public Health, 1870–1920." *Journal of the History of Medicine and Allied Sciences* 29, no. 1: 55–73.

———. 1985. "The Search for Professional Order in 19th-Century American Medicine." In *Sickness and Health in America*, edited by Judith Walzer Leavitt and Ronald L. Numbers, 219–232. Madison: University of Wisconsin Press.

Rosner, David. 1982. *A Once Charitable Enterprise: Hospitals and Health Care in Brooklyn and New York, 1885–1915*. New York: Cambridge University Press.

Rothman, David J. 1991. *Strangers at the Bedside: A History of How Law and Bioethics Transformed Medical Decision Making*. New York: Basic Books.

Rothstein, William G. 1992. *American Physicians in the Nineteenth Century: From Sects to Science*. Baltimore, MD: Johns Hopkins University Press.

Rudolph, Frederick, and John R. Thelin. 1991. *The American College and University: A History*. Athens: University of Georgia Press.

Rutkow, Ira. 2010. *Seeking the Cure: A History of Medicine in America*. New York: Scribner.

Saks, Mike. 2003. *Orthodox and Alternative Medicine: Politics, Professionalization, and Health Care*. New York: Continuum.

Scheid, Volker. 2002. *Chinese Medicine in Contemporary China: Plurality and Synthesis*. Durham, NC: Duke University Press.

Scott, James C. 1985. *Weapons of the Weak*: New Haven, CT: Yale University Press.

Scott, W. Richard, Martin Ruef, Peter J. Mendel, and Carol A. Caronna. 2000. *In-*

stitutional Change and Healthcare Organizations: From Professional Dominance to Managed Care. Chicago: University of Chicago Press.

Schudson, Michael. 1981. *Discovering the News: A Social History of American Newspapers*. New York: Basic Books.

———. 1998. *The Good Citizen: A History of American Civic Life*. New York: Martin Kessler Books.

Schweber, Libby. 2006. *Disciplining Statistics: Demography and Vital Statistics in France and England, 1830–1885*. Durham, NC: Duke University Press.

Sewell Jr., William. 1992. "A Theory of Structure: Duality, Agency, and Transformation." *American Journal of Sociology* 98, no. 1: 1–29.

———. 2005. "The Concept(s) of Culture." In *Practicing History: New Directions in Historical Writing after the Linguistic Turn*, edited by Gabrielle M. Spiegel, 141–161. New York: Psychology Press.

Shapin, Steven. 1994. *A Social History of Truth: Civility and Science in Seventeenth-Century England*. Chicago: University of Chicago Press.

———. 2008. *The Scientific Life: A Moral History of a Late Modern Vocation*. Chicago: University of Chicago Press.

Shapin, Steven, and Simon Schaffer. 1985. *Leviathan and the Air-Pump: Hobbes, Boyle, and the Experimental Life*. Princeton, NJ: Princeton University Press.

Shorter, Edward. 1985. *Bedside Manners: The Troubled History of Doctors and Patients*. New York: Simon and Schuster.

Shortt, S. E. D. 1983. "Physicians, Science, and Status—Issues in the Professionalization of Anglo-American Medicine in the 19th-Century." *Medical History* 27, no. 1: 51–68.

Shryock, Richard H. 1967. *Medical Licensing in America, 1650–1965*. Baltimore, MD: Johns Hopkins University Press.

Sica, Alan. 2010. "Merton, Mannheim, and the Sociology of Knowledge." In *Robert K. Merton: Sociology of Science and Sociology as Science*, edited by C. Calhoun, 164–182. New York: Columbia University Press.

Silverman, Kenneth. 1984. *The Life and Times of Cotton Mather*. New York: HarperCollins.

Smillie, W.G. 1943. "The National Board of Health 1879–1883." *American Journal of Public Health* 33, no. 8: 925–930.

Spink, Wesley. 1978. *Infectious Diseases: Prevention and Treatment in the Nineteenth and Twentieth Centuries*. Minneapolis: University of Minnesota Press.

Starr, Paul. 1976. "The Politics of Therapeutic Nihilism." *Hastings Center Report* 6, no. 5: 24–30.

———. 1982. *The Social Transformation of American Medicine*. New York: Basic Books.

———. 1987. "The Sociology of Official Statistics," In *The Politics of Statistics*, edited by William Alonso and Paul Starr, 7–58. New York: Russell Sage Foundation.

———. 2011. *Remedy and Reaction: The Peculiar American Struggle over Health Care Reform*. New Haven, CT: Yale University Press.

Stevens, Rosemary. 1971. *American Medicine and the Public Interest*. New Haven, CT: Yale University Press.

Stevenson, Lloyd G. 1965. "Putting Disease on the Map: The Early Use of Spot Maps in the Study of Yellow Fever." *Journal of the History of Medicine and Allied Sciences* 20, no. 3: 226–261.

Strong, Philip. 1990. "Epidemic Psychology: A Model." *Sociology of Health and Illness* 12, no. 3: 249–259.

Susser, Mervyn, and Ezra Susser. 1996. "Choosing a Future for Epidemiology: I. Eras and Paradigms." *American Journal of Public Health* 86, no. 5: 668–668.

Temkin, Owsei. 1977. *The Double Face of Janus and Other Essays in the History of Medicine*. Baltimore, MD: Johns Hopkins University Press.

Tesh, Sylvia Noble. 1988. *Hidden Arguments: Political Ideology and Disease Prevention Policy*. New Brunswick, NJ: Rutgers University Press.

Thompson, E. P. 1968. *The Making of the English Working Class*. London: IICA.

Tilly, Charles. 2006. *Why? What Happens When People Give Reasons . . . And Why*. Princeton, NJ: Princeton University Press.

Timmermans, Stefan, and Hyeyoung Oh. 2010. "The Continued Social Transformation of the Medical Profession." *Journal of Health and Social Behavior* 51, no. 1 suppl.: S94–S106.

Tomes, Nancy J. 1997. "American Attitudes toward the Germ Theory of Disease: Phyllis Allen Richmond Revisited." *Journal of the History of Medicine and Allied Sciences* 52, no. 1: 17–50.

———. 1998. *The Gospel of Germs: Men, Women, and the Microbe in American Life*. Cambridge, MA: Harvard University Press.

Tomes, Nancy J., and John Harley Warner. 1997. "Introduction to Special Issue on Rethinking the Reception of the Germ Theory of Disease: Comparative Perspectives." *Journal of the History of Medicine and Allied Sciences* 52, no. 1: 7–16.

Toulmin, Stephen. 1992. *Cosmopolis: The Hidden Agenda of Modernity*. Chicago: University of Chicago Press.

Trachtenberg, Alan. 2007. *The Incorporation of America: Culture and Society in the Gilded Age*. New York: Hill and Wang.

Van Ingen, Philip. 1949. *The New York Academy of Medicine: Its First Hundred Years*. New York: Columbia University Press.

Vaughan, Diane. 1996. *The Challenger Launch Decision: Risky Technology, Culture, and Deviance at NASA*. Chicago: University of Chicago Press.

———. 1999. "The Role of the Organizations in the Production of Techno-Scientific Knowledge." *Social Studies of Science* 29, no. 6: 913–943.

———. 2004. "Theorizing Disaster: Analogy, Historical Ethnography, and the Challenger Accident." *Ethnography* 5, no. 3: 315–347.

Vernon, Keith. 1990. "Pus, Sewage, Beer and Milk—Microbiology in Britain, 1870–1940." *History of Science* 28, part 3, no. 81: 289–325.

Veysey, Laurence R. 1965. *The Emergence of the American University*. Chicago: University of Chicago.

Vinten-Johansen, Peter, Howard Brody, Nigel Paneth, Stephen Rachman, and Michael Russell Rip. 2003. *Cholera, Chloroform, and the Science of Medicine: A Life of John Snow*. New York: Oxford University Press.

Waldor, Matthew K, Eric J. Rubin, Gregory D. N. Pearson, Harvey Kimsey, and John J. Mekalanos. 2003. "Regulation, Replication, and Integration Functions of the Vibrio Cholerae CTX Are Encoded by Region RS2." *Molecular Microbiology*, 24, no. 5: 917–926.

Warner, John Harley. 1991. "Ideals of Science and Their Discontents in Late 19th-Century American Medicine." *Isis* 82, no. 313: 454–478.

———. 1997. *The Therapeutic Perspective: Medical Practice, Knowledge, and Identity in America, 1820–1885*. Princeton, NJ: Princeton University Press.

———. 1998. *Against the Spirit of System: The French Impulse in Nineteenth-Century American Medicine*. Princeton, NJ: Princeton University Press.

———2002. "The Fall and Rise of Professional Mystery: Epistemology, Authority, and the Emergence of Laboratory Medicine in Nineteenth-Century America." In *The Laboratory Revolution in Medicine*, edited by Andrew Cunningham and Perry Williams, 110–141. New York: Cambridge University Press.

Weber, Max. 2002. *The Protestant Ethic and the "Spirit" of Capitalism and Other Writings*. New York: Penguin Books.

Weick, Karl E. 1979. *The Social Psychology of Organizing*. Reading, MA: Addison-Wesley.

Weisz, George. 1978. "The Politics of Medical Professionalization in France 1845–1848." *Journal of Social History* 12, no. 1: 3–30.

Wheatley, Steven C. 1988. *The Politics of Philanthropy: Abraham Flexner and Medical Education*. Madison: University of Wisconsin Press.

White, Hayden. 1980. "The Value of Narrativity in the Representation of Reality." *Critical Inquiry* 7, no. 1: 5–27.

Whooley, Owen. 2008. "Objectivity and Its Discontents: Knowledge Advocacy in the Sally Hemings Controversy." *Social Forces* 86, no. 4: 1367–1389.

———. 2010. "Diagnostic Ambivalence: Psychiatric Workarounds and the Diagnostic and Statistical Manual of Mental Disorders." *Sociology of Health and Illness* 32, no. 3: 452–469.

Whorton, James C. 1982. *Crusaders for Fitness: The History of American Health Reformers*. Princeton, NJ: Princeton University Press.

———. 2002. *Nature Cures: The History of Alternative Medicine in America*. New York: Oxford University Press.

Wilson, Major L. 1974. *Space, Time, and Freedom: The Quest for Nationality and the Irrepressible Conflict, 1815–1861*. Westport, CT: Greenwood Press.

Worboys, Michael. 2000. *Spreading Germs: Diseases, Theories, and Medical Practice in Britain, 1865–1900*. New York: Cambridge University Press.

Young, James Harvey. 1967. *The Medical Messiahs: A Social History of Health Quackery in Twentieth-Century America*. Princeton, NJ: Princeton University Press.

Zerubavel, Eviatar. 1999. *Social Mindscapes: An Invitation to Cognitive Sociology*. Cambridge, MA: Harvard University Press, 1999.

———. 2003. *Terra Cognita: The Mental Discovery of America*. New Brunswick, NJ: Transaction Publishers.

———. 2011. *Ancestors and Relatives: Genealogy, Identity, and Community*. New York: Oxford University Press.

Ziporyn, Terra. 1988. *Disease in the Popular American Press: The Case of Diphtheria, Typhoid Fever, and Syphilis, 1870–1920*. New York: Greenwood Press.

INDEX

Note: Bold page numbers refer to figures and tables.